Red Tourism in China

I0131791

This book analyzes the phenomenally profitable "Red Tourism" industry in China, in which visitors make pilgrimages to sites of historical significance to the Communist Party of China and the Chinese Revolution.

The book examines Red Tourism in connection with the transforming power relations between the state and the private, communication in the socialist past, and the current round of capitalization, against the backdrop of the world's second largest economy. By re-evaluating the conventional notion of propaganda through the lens of neutral *xuanchuan* propaganda, the book presents a nuanced look at the social space of Red Tourism, revealing that propaganda should be conceived as a commodity, an industry, or even a media system similar to the news media.

Drawn from combining fieldwork and cultural analysis spanning a decade, this book will be of interest to students and scholars of communication studies, tourism, and Chinese politics.

Chunfeng Lin is Associate Professor of Communication at East China Normal University, China.

Routledge Contemporary China Series

For more information about this series, please visit: https://www.routledge.com/
Routledge-Contemporary-China-Series/book-series/SE0768

Red Tourism in China
Commodification of Propaganda

Chunfeng Lin

Routledge
Taylor & Francis Group

LONDON AND NEW YORK

First published 2023
by Routledge
4 Park Square, Milton Park, Abingdon, Oxon OX14 4RN

and by Routledge
605 Third Avenue, New York, NY 10158

Routledge is an imprint of the Taylor & Francis Group, an informa business

© 2023 Chunfeng Lin

The right of Chunfeng Lin to be identified as author of this work has been
asserted in accordance with sections 77 and 78 of the Copyright, Designs
and Patents Act 1988.

All rights reserved. No part of this book may be reprinted or reproduced or
utilised in any form or by any electronic, mechanical, or other means, now
known or hereafter invented, including photocopying and recording, or in
any information storage or retrieval system, without permission in writing
from the publishers.

Trademark notice: Product or corporate names may be trademarks or
registered trademarks, and are used only for identification and explanation
without intent to infringe.

British Library Cataloguing-in-Publication Data
A catalogue record for this book is available from the British Library

Library of Congress Cataloging-in-Publication Data
A catalog record has been requested for this book

ISBN: 978-1-032-13960-9 (hbk)
ISBN: 978-1-032-13981-4 (pbk)
ISBN: 978-1-003-23178-3 (ebk)

DOI: 10.4324/9781003231783

Typeset in Times New Roman
by Deanta Global Publishing Services, Chennai, India

Contents

Illustrations

Figures

Table

Acknowledgments

This book is based on my dissertation at the Institute of Communications Research (ICR) at the University of Illinois at Urbana-Champaign, my alma mater from 2011 to 2018. My deepest gratitude goes to my doctoral advisor John Nerone, who not only inspired me with his critique of the hegemonic western model of journalism, but also dispelled my frustration with his dark humor. Him introducing me as his "last advisee" to the academic world was indescribably heartwarming and made me truly proud. I would also like to thank my doctoral committee: Poshek Fu, Bob McChesney, James Hay, and Cameron McCarthy. I likened this committee to as an ensemble of da Vinci, Picasso, and Monet for each of them is a distinguished scholar from their distinctive field. They generously offered their enthusiasm and expertise as the project took shape. This book would not have been possible without their mentorship.

I wrote this book over a decade throughout my stay in the United States and China, and owe thanks to many people who, knowingly and unknowingly, helped bring this book into existence. Among the institutions that supported this project through research funding are the Graduate College at Illinois, which awarded me a generous Dissertation Completion Fellowship, and the East China Normal University, my current post. Several professors during my days at Illinois were especially important mentors. Dan Schiller, Norman Denzin, Clifford Christians, Kent Ono, Rich Martin, CL Cole, Ann Reisner, and above all Angharad Valdivia deserve special mention. It brings me joy to note that I have sat at kitchen tables with many aforementioned. These academic exchanges have often sparked passion and inspiration for my academic journey. The evolution of this book owes a personal debt to my Illinois cohorts.

Routledge has been supportive and efficient. Stephanie Rogers saw the value of this project and worked diligently to bring it to fruition. Emily Pickthall and Andrew Leach assisted me with the preparation of the manuscript. All of them have earned my gratitude. I also thank three anonymous readers, whose comments have made this a better book. Parts of Chapter 2 and Chapter 5 have appeared in *Asian Journal of Communication* (Issue 5, 2017) and *Communication and Critical/Cultural Studies* (Issue 3, 2015). I thank the publishers of the journals for permission to reproduce copyrighted material.

Family members have greatly contributed to the book. Chunchen Lin has always been supportive in hosting and touring me across San Jose, California. Linhai Zhu, an Illinois engineering graduate, helped proofread portions of the manuscript and made suggestions to make it more accessible to lay persons. It would have been a very different book without his input.

Finally, I am indebted to my father, Yushan Lin, who passed away a few months after I returned to China from the United States. He asked me to write 300 words in a journal every day when I was a toddler. I have now realized how vital this once unappealing task has prepared me for writing throughout my journalistic and academic career. He was among the first generation of people in New China to hold a bachelor's degree in journalism from one of the most prestigious universities in the country. My father was a journalist, propagandist, and journalism professor. It is precisely his life's journey that has helped me view propaganda as a social space. He was the best teacher of all.

Chunfeng Lin
Shanghai, 2022

1   Introduction

A solo elderly female backpacker was asking the bus driver where she could take public transportation next morning to Yucha Gorge, a tourist spot in the vicinity of Yan'an city, China. The driver was very kind, detailing all possible routes along with other information. Casual inquiry turned quickly into a friendly chat. It turned out that the backpacker had visited Yan'an several times. In addition to the Red Tourism sites she visited earlier, she would like to explore some new places. Beyond this motive, the *bona fide* reason she came back to Yan'an was that she found the flights from Shanghai to Yan'an incredibly cheap. This was part of the deal of a new initiative of the city of Yan'an entitled, "Red Stars Flying Me to Yan'an from Thirteen Cities." Aiming at promoting the revolutionary history, the project was significant not only because it was during the middle of the COVID-19 pandemic, but also because it was launched right before the centenary anniversary of the founding of the Communist Party of China (CPC).

This vignette illustrates a series of contradictions rooted in Red Tourism, historical seriousness and touristic fun, strictly controlled project of the state, and considerable latitude of individual tourists' itineraries to name a few. I saw the backpacker on June 9, 2021, a few days before the Dragon Boat Festival, a three-day public holiday in China. The day was special to me for it marks the ten-year anniversary of my Red Tourism research on which this book is based. Too much has happened in the past decade. When this project started in 2011, for example, Yan'an Nanniwan Airport was merely a blueprint. The airport has now been erected from undeveloped land.

There are much more. Around this project's start, two principal notions of the book – the "Red Tourism industry" and the "commodification of propaganda" – were merely theoretical constructs, inviting intense debates. In 2021, the well-organized and highly productive Red Tourism industry is discernable in many places and as does the commodification of propaganda. In Yan'an, the primary fieldwork site of this book, the massive and controversial development project "Golden Yan'an" has become a reality by 2021. Part of the Holy Land Valley Cultural Tourism Industrial Park, Golden Yan'an is a town built from scratch. Featuring old fashioned Chinese streets and shops of Yan'an in the Shaanxi-Gansu-Ningxia border region in the 1930s, the Red-themed, tens of billions of yuan tourist town is now branded as a "new landmark of Red Tourism in China"

DOI: 10.4324/9781003231783-1

(Chao, 2013). Each evening, large groups of tourists who have been traveling all day in other Red Tourism sites of the city are ferried to this site via tour buses one after another to stay the night and enjoy the city's amenities.

While the attraction of Golden Yan'an is of nostalgia, the selling point of the nightlight show of "The Ode of Yan'an" is cutting-edge technology. The nightly half-hour performance at 8:30 pm along the Yanhe River with free admission is absolutely sensational. It features programmed lighting effects on the three iconic mountains of Yan'an and an audio-visual show projected on Baota Mountain, the landmark of the city, across 15,000 square meters. As the title suggests, the show glorifies the Yan'an Period of the party by representing a series of historical moments of the time. The company in charge of the show has also contracted with multiple international events such as the Beijing Olympic Games, Hangzhou Corporation Convention, etc. Meanwhile, this free nightlight show has turned the city of Yan'an into a popular hotspot.

All these contradictions and rapid developments concerning Red Tourism encapsulate the very essence of this book.

Red Tourism (*hongse lvyou*) refers to a form of tourism of the People's Republic of China in which visitors make pilgrimages to sites of historical significance to either the Communist Party of China (CPC) or the Chinese Revolution. The profitability of this so-called "Red" economy has been phenomenal: according to The *People's Daily*, "last year [2011] alone, China's Red Tourism sites received 540 million visitors, counting for 20 percent of tourists in the country" (Jin, 2012). According to a governmental outline from 2011, by 2015 the national Red tourist travel number would exceed 800 million with an average annual growth of 15 percent, and would eventually grow to a quarter of total domestic tourism in China; the consolidated revenue of Red Tourism would reach 200 billion yuan (US$31 billion) with an average annual growth rate of 10 percent (General Office, 2011). China's Red Tourism wave had even reached out internationally. To accommodate the burning enthusiasm of Chinese Red tourists in Russia, Russian tourism authorities launched a special route called the "red circuit," an eight-day tour across multiple Russian cities featuring the life trajectory of Vladimir Lenin (Koreneva, 2015).

A heritage tourism, Red Tourism is a kind of oxymoron. It is a yoking together of two extraordinarily powerful drivers in the cultural language surrounding China. On the one hand, "Red" is associated with ideology, discipline, and loyalty combined with a history of authentic struggle and liberation. On the other hand, tourism designates an ongoing process of capitalization. Eerily, this oxymoronicality of Red Tourism epitomizes the cultural imagination of contemporary China, in which the Chinese state seems simultaneously to be engaged in mining the past for its lucrative images and narrative resources as well as calculating a future linked to a kind of Red economy where propaganda roots, morphs, and thrives. Nevertheless, this is rarely understood, in part because the way we see propaganda practice has been fixed; scholars treat propaganda as a method, a technique, and/or a practice, devoting much attention to journalism, the news industry, the media system, and censorship while devoting less attention to the social space of propaganda. This research is meant to fill that gap.

That said, this is not a typical tourism study project. It treats tourism as a mediated space, a media system in which participating tourists act like media as they distribute, circulate, and propagate intended messages so similar to mass media. In that sense, this is a media study project *per se*.

This study specifically examines Red Tourism in connection with the transforming power relations between the state and the private, communication in the socialist past and the current round of capitalization, against the backdrop of the world's second largest economy. It argues that Red Tourism is a social space comprising all sorts of political pathways and communication networks. As such, Red Tourism is both produced and productive: it was produced by physical space such as cities, tourist sites, and museums, and also abstract space including histories, ideologies, collective memories, (post)socialist nostalgia, social imaginaries, and so forth. At the same time, it is (re)producing all of these physical and non-physical networks by following distinctive mechanisms combining the dominant ideology and the capitalist mode of production. It therefore calls for a new theoretical framework and approach that does not belong to conventional research in propaganda.

I adopted an interpretive approach invoked by two strands of Marxism that speak at times at cross purposes regarding the production of meanings and power relations. In the first strand, that of the contemporary Western Marxism of Henri Lefebvre and Raymond Williams, there is a sophisticated reading of the production of superstructure not as a static, determined process but through dynamic mediation. They help rethink Red Tourism not as simply a propaganda project determined by the Party's ideology but a social space mediated by and mediating communications. The second strand, political economy, derived from classical Marxism, carries out inquiries into the relation, production, distribution, and exchange of power in the capitalist mode of production. It is a useful tool to analyze the political-economic base of Red Tourism and related issues. Therefore, in this study I pursued what Robert Babe (2009) refers to as "a dialectical middle ground," integrating critical/cultural studies and political economy of communication (p. 5). It is also what Mattelart (2008) refers to as a "cross-disciplinary approach" combining the discursive analysis with the political economy of communication (p. 25).

This book sits at the intersection of media studies, critical studies of tourism, and political economy of communication in addition to being a part of a series of studies on contemporary China. Each of the three areas engages with a specific dominant topic that can be reduced to a keyword. The three keywords are media, space, and commodification. They serve as both the cornerstones of the theoretical construct of the book and the conjunctions of the two strands of scholarship aforementioned. Note that by media I mean both tourism and conventional types of media. Not only are the three elements fundamental to Red Tourism; they are also critical to a profound understanding of the Chinese state and its communication. In other words, they are decisive and inseparable in mapping out the ongoing transformation of the geography of contemporary Chinese mass communication. Red Tourism connects the critical components of Chinese communication and converts them into an exceptional case that is visually comprehensible and accessible to readers. Think of Mao Zedong's portrait hung at Tiananmen Gate: it immediately legitimates the history of the nation-state through spatial articulation. Then think about Red Tourism as a result of a massive propagation of such symbolism all over the country through, among others, the process of commodification. Now propaganda and capital overlap. Together, they form the complex and intriguing reality of Red Tourism.

One way of making sense of the current study is to introduce the three keywords and their meanings in specific contexts. Since this work deals with highly ambiguous and, perhaps, forever debatable subjects in communication study including media, propaganda, political tourism, social space, and commodification in a non-English-speaking, non-Western-democratic country, I found this strategy more appropriate for the introduction. Nevertheless, this is not done in a pick-and-choose manner: reiterating varied definitions of each term and taking one that fits. Rather, I want to examine different ideas surrounding the keywords and to answer a few critical questions in much wider social and cultural contexts, and ultimately, to make my own case. Those questions are: why did I choose *xuanchuan* 宣传, the Chinese term for propaganda? How do I see tourism? Is Marxism applicable to China's case? In doing so, I also attempt to furnish proper theoretical grounds on which this study is based, constructed, and developed. In a sense, I envisage the following discussion of the three keywords to serve as a simplified tourist map to acquaint the readers with the landmarks, the travel routes, and the destinations, and at the same time, to show how I planned to get there. This is to say, in what follows, I want to offer both a "map" and a "tour" of the book.

Media: why *xuanchuan*?

Instead of getting straight to the point, this book begins with a theoretical exploration of the Chinese term for propaganda. I want to offer a new theoretical framework of propaganda, which I refer to as the "*xuanchuan* model." This is a bold yet necessary attempt in several regards.

Propaganda is *the* great myth in communication studies. On the one hand, propaganda study is the birthplace where modern communication study as a field developed and thrived; it generated the backbone theories of communication, particularly in media effect research (e.g., strong, limited, and non-effect theories) as well as enormous institutional power. On the other hand, contemporary propaganda research, as a strand of communication study, empirical or theoretical, seems to have been exhausted. Jowett and O'Donnell (2012, xiii) have noted that "[p]ropaganda has received scant attention as a subject in its own right within the spectrum of communication studies." By contrast, "propaganda," the term, has increasingly gained visibility in popular discourse and polemics. Finally, we realize that it is even impossible to locate a commonly accepted definition of the term. This is to say, what is really mythical about propaganda is that although we think we know a whole lot about propaganda in terms of history, methods, techniques, and practices, we barely know what it is. For example, what is the difference between propaganda and, say, ideology, public relations, and persuasion? For many, it seems that the only difference lies in propaganda's negative connotation.

But this is untrue. The theoretical exploration of *xuanchuan* is meant to disturb the dominance of the Western notion of propaganda in which propaganda, whatever defined, is believed to be a nefarious thing, loosely associated with brainwashing. Socolow (2007), for example, remarked that the word "propaganda" connotes "conspiratorial and anti-democratic" in US politics (p. 109). Cunningham (1992) also pointed out that propaganda in English-speaking countries is an "ominous term" and "fraught with intentional deception" (p. 233). Although in the West a neutral thesis about propaganda can be heard over time from Bernays (1928; 1942), LaPiere and Farnsworth (1936), Lasswell (1948), Ellul (1965), Welch (1983), and Jowett and O'Donnell (2012), the negative connotation of propaganda has gained its hegemonic power all over the world. The hegemonic power of the Western model of propaganda renders the study of Red Tourism less meaningful. Here is an anecdote. I was repeatedly asked by the academics on many occasions where I presented or talked about my research on Red Tourism, whether or not I would consider Red Tourism propaganda. The hidden assumption goes like this: if it is propaganda, then everything about Red Tourism goes to the dark connotation of propaganda; if not, then there is no need to view Red Tourism as such. The word "propaganda" itself is deterministic, in other words. Hence, searching for a neutral framework about propaganda becomes imperative, and *xuanchuan* is the one I offer.

Xuanchuan has different meanings. In Chinese history, culture, and social practice, *xuanchuan*/propaganda is not a pejorative term. *Xuanchuan* generally means "disseminating purposeful information" and traditionally is associated with good things and deeds (Lin, 2015; Lin & Nerone, 2015). In China, news coverage oftentimes is called "news propaganda (*xinwen xuanchuan*)". This is because, as we already know, news is socially constructed and meant to deliver certain, thusly, "purposeful" information to the public. Ideally, journalism is good for the general public. For this reason, mainland Chinese journalists sometimes call themselves "the people doing propaganda" (*gao xuanchuan de*) in a rather humble tone. Liang Qichao (2010), arguably the most distinguished Chinese liberal scholar in the late Qing and early Republican periods, proclaimed himself an "aggressive propagandist" (p. 99). The examples can be multiplied. The point is that they point to another universe of propaganda where propaganda is not necessarily evil. Nevertheless, I do not mean to imply by this that *xuanchuan* is one thing, good, and propaganda another, and bad. There is a cultural gap between *xuanchuan* and propaganda but they are complementary to each other and situated in different contexts.

To situate my study in the context of Chinese communication, I adopted the Chinese neutral connotation of *xuanchuan*/propaganda. There is no Chinese term for either "good propaganda" or "bad propaganda." It would be less confusing to non-Chinese speakers otherwise. According to the *Modern Chinese Dictionary*, *xuanchuan* has a range of meanings. *Xuanchuan* can mean "announce" and "convey" as it is in a common Chinese phrase "*xuanchuan zhengce*" (publicizing policies). *Xuanchuan* can also refer to "explaining to and educating the people." Closer to the Western notion of propaganda is *xuanchuan*'s meaning of propagation or spread of information. Again, no derogatory sense attached. What I mean by propaganda/*xuanchuan* in this study is a combination of all these meanings. This is to say, referring to Red Tourism as propaganda does not automatically mean that I consider Red Tourism deceptive and thusly a lie. As earlier mentioned, Red Tourism represents an authentic revolutionary history, though mediated. Red Tourism is propaganda/*xuanchuan* only because it is conveying historical narratives, publicizing the revolutionary history, educating tourists, and spreading out intended information and ideology.

Space: how do I see tourism?

Upon introducing the neutral propaganda framework, I then delve into the cultural roots of Red Tourism. I will try to sketch out a prehistory and a history of Red Tourism, which is about unfolding untold histories of political tourism in China. By "political tourism" I simply mean journeys with a political purpose, explicitly or implicitly. In China, political tourism may include, for example, politically charged and motivated journeys made by *youshi* 游士, or Chinese lobbyists, in earlier times, journeys made in response to the call of Mao Zhedong during the Cultural Revolution, or spring trips to revolutionary martyrs' cemeteries arranged by schools during both socialist and post-socialist China, and so forth. This suggests that the political signification can come from travelers, destinations, or organizers. Note that this is not unique to Red Tourism. Tourism, by and large, is political. I will illustrate this perspective in Chapter 3.

Tourism has two common and interlinked meanings. Tourism can either refer to a practice of traveling for pleasure or the business of providing such services. Vernacularly tourism has been viewed as a recreation or leisure activity. But contemporary critical studies of tourism tend to problematize this approach. Within these theories – whether Dean MacCannell (1999) stressing the interplay between tourists' behaviors and social relations or John Urry (2002) revealing the interwoven relationship of gazing and theming – tourism has been deemed as a site of power, whether in the form of "staged authenticity" or "high levels of surveillance." Departing from here and following this strand of critical scholarship, in this book I treat tourism as a media system, somewhat like the press. Let me explain.

Like the press, a central role of tourism is to spread out information to the common people. Tourism can be a shaper like the press in forming and shaping public opinion. In addition, like journalism, tourism also matters to "public intelligence," to borrow a term from Nerone (2015). Specifically, tourism informs visitors by presenting tourists with information regarding local culture, nature, arts, history, politics, and so forth. This is in part what journalism sets out to do. What's more, tourism is an "ism" too. To put it another way, tourism is, too, a site of ideology. This is a deeper underlying meaning of tourism. The ideology glued to the touristic spot is readily to be grasped by tourists. A colonial plantation or a Disney theme park is somewhat like a newspaper, so much so that whether news articles, news photos, and commentaries in the case of newspaper, or historical buildings, plaques, and emblems in the case of the plantation, and designs and settings in the case of a theme park, all deliver intended messages to the readers or visitors. Such messages are by no means value-free. They are ideological, and potentially political. Messages delivered in touristic sites are carefully selected, coded, framed, filtered, and organized through spatial articulations along with other mechanisms. In *Searching for Yellowstone*, Denzin (2008) captures racial and gender stereotypes, among other things, through a critical reading of the first American national park. Jean Baudrillard (1994) viewing Disneyland as a simulacrum in which American ideology is represented, is another case in point.

Tourism is a media system just like the news media. Tourism has its own institutions (e.g., central and local governments, tourism bureaus/authorities) and systems and mechanisms of production (e.g., travel agencies, tourism companies), circulation (e.g., tourist routes, boundaries, maps), and even censorship (e.g., passports, visas, international relations, restrictions, and customs authorities). Messages and ideologies travel with tourists traveling across space. And information, too, travels unevenly. Like the Western countries dominating the international news flow (Kim & Barnett, 1996), the inequality of international tourists' flow is evident. A home to a large variety of wild animals, South Africa is a popular tourist destination for Europeans. Rarely do local South African residents have equal agency to travel in Europe for leisure. This leads MacCannell (1999) to argue that tourism is a better structure than nation when accessing economic and social difference across international borders and cultural boundaries.

Beyond the functional analogy between journalism and tourism is a connecting point: public opinion. For Nerone (2015), journalism can be understood as the representation of public opinion through the journalist's proxy role for the public. So does tourism. Tourism holds power accountable to something some would call public opinion. But it is the people who run things who decide on behalf of the public which places are to be developed and open to the public and which not. This mechanism parallels that of journalism: news messages are filtered and manufactured to be "representative" of public opinion (Herman & Chomsky, 1988/2008). *The New York Times*' famous slogan is "All the News That's Fit to Print." However, determining whether a news report is "fit" or not is partly an art of politics and partly a maneuver in economics. In other words, it is the combination of money and power that decides such fitness (Herman & Chomsky, 1988/2008). Like those "fitted" news articles published in the mainstream news outlets in the United States, the Statue of Liberty, the Empire State Building, the Metropolitan Museum of Art, and Times Square along many other sites all perfectly "fit" in New York tourism. Slums, on the other hand, would never do the trick. While some unpleasant places such as slaughterhouses, sewers, and the morgue can be tourist attractions as examined by MacCannell (1999), those were repurposed rather to represent a moment of industrialization and modernization than to get a good grasp of the reality.

Like reading news, tourists can only tour the places that are intended to be seen. Like journalism, tourist sites, too, inform the general public on a wide range of critical matters about state, nation, history, race, culture, gender, politics, and more. Mount Rushmore National Memorial of South Dakota is a notable example. Featuring sculptures of the heads of four United States presidents, Mount Rushmore has become an iconic symbol of the nation. But the ideological signification of this space is fiendishly complicated, particularly considering that the United States plundered the place from Native Americans in the late 19th century. This is to say, for Native Americans Mount Rushmore signifies the shocking savagery of their ethnic history. As a competing monument, the Crazy Horse Memorial is under construction, which reminds one of the fights of Native Americans against US federal government encroachments. But one might wonder, how many tourists would choose visiting the Crazy Horse over Mount Rushmore. The answer is not encouraging. In journalism studies, theoretical constructs like "agenda setting," "framing," and "the propaganda model" suggest a similar effect that what we get from news is rather a matter of what the news media feed us. It is not an exaggeration that every "ism" in journalism can be found in tourism.

Unlike journalism, however, messages delivered in tourism are not dominantly through verbal language, but mostly through spatial articulation. Semiotic scholarship is of paramount importance in this regard. Regarding architecture as mass communication, for example, Eco (1997) has pointed out that architectural objects could connote a certain ideology through what he calls a "code-language." It is an architectural code system, including "technical codes (dealing with architectural engineering)," "syntactic codes (concerning spatial types)," and "semantic codes (concerning the significant units of architecture)" (p. 193). But space in this study is more than this.

Why social space?

The neutral theoretical framework of *xuanchuan*/propaganda intersects with histories of Red Tourism at the point that propaganda in China is not merely a technique, ideology, action, but something rather fuzzy, including all of them, and much more. It is on this basis that I applied Lefebvre's (1991) seminal idea of social space to the case of China's propaganda. Through a spatial analysis of Red Tourism in relation to lived experience of the Chinese people, I revealed a social space of propaganda, far more nuanced and sophisticated than Lasswell, Lazarsfeld, and Ellul – three great authors who write about propaganda – could have imagined.

In this book, space can be understood at two levels. Macroscopically, space is pointing to a spatial perspective, a particular way of thinking about communication as a spatial project. Instead of treating Red Tourism as a temporal project, and, correspondingly, looking for its origins, developments, changes, and effects, this study treats Red Tourism as a spatial project, consisting of maps, places, sites, routes, tourists, and the like. However novel it may sound, thinking communication as space is relatively "old." Communication as a field started with scholars imagining something bigger than merely a linear information delivering process as suggested by Information Theory, Claude Shannon's *magnum opus*. Sharing origins with "commune," "communion," and "community," communication was once perceived as a spatial project, involving all kinds of networks across all realms. Those networks include national markets and international trading routes in the economic realm (Mattelart, 2000), the public sphere in the social realm (Habermas, 1989), and the Sacred Congregation for Propagating the Faith of the Roman Catholic Church in the spiritual realm (Mattelart, 1996), just to name a few. This is to say, communication was believed to be produced by and attached to space across all social realms. Many scholars have already said that, implicitly if not explicitly. A quick catalog of the influential authors who have connected communication to space includes Henri Lefebvre, Benedict Anderson, Armand Mattelart, Umberto Eco, James Carey, David Harvey, Walter Benjamin, Jürgen Habermas, Georg Simmel, Sigfried Kracauer, John Hannigan, and many others. One of the benefits of regarding communication as space from these authors is to locate communication at a blueprint of human society, which, in turn, helps extend our knowledge on critical issues such as nationalism (Anderson, 2006), modernity (Harvey, 2003), urban lifestyle (Kracauer, 1975), democracy (Habermas, 1989), and economy (Hannigan, 1998).

Microscopically, and more importantly, space refers to social space in this study. It is indeed one of the major theoretical constructs of this book. Lefebvre (1991) noted that "social space is not *a* thing" (p. 73, emphasis mine). It is everything, physical and non-physical, within the space. As such, a social space is both produced and producing. Coincident with the subject matter of tourism of this book, Lefebvre illustrated what he called the "polyvalence of social space" by examples of a series of Italian popular cities for tourists. In doing so, Lefebvre does more than simply account for how those cities, as social space, were produced and what they are producing. Lefebvre (1991) called attention to the interconnection between the two processes, saying, "[m]ediations, and mediators, have to be taken into consideration: the action of groups, factors within knowledge, within ideology, or within the domain of representations" (p. 77).

Lefebvre's notion of space helps rethink Marx's theory about the relationship between the base and superstructure. It is a popular though superficial reading of Marx's axiom as a combination of a determining base and a determined superstructure. Strictly following such "determinism" would make the task of analyzing culture not only mechanically boring but also potentially misleading. For one thing, culture is an accumulative process across histories that cannot be simply determined by a given and fixed base. For another, the discovery of cultural patterns requires us to examine what Williams (1961/1975) calls "structures of feeling," some kind of living experience of the people across different generations. Perhaps the determining relationship was not what Marx meant to tell us. Like Lefebvre, critical theorist Raymond Williams, among others, has also challenged the dominant reading. Williams (1980/1997; 2001/2006) remarked that two qualifications about the relationship between the base and superstructure have to be addressed: the "famous lags," which denote the temporal distance between certain economic activities and their corresponding cultural activities, and the notion of "mediation" in which the superstructure does not "reflect" but mediate the base. Simply put, it is a process of mediation. Williams (1980/1997) argued that the base is not an abstract "state" but a dynamic "process" (p. 34).

Back to the current study, one of the central arguments of the book is that as a social space Red Tourism is not merely produced by a particular history, ideology, and economy, and determined by them; it is also productive, producing values, imaginations, and a so-called "Red industry" (*hongse chanye*). The capitalism involved in the production of this social space, particularly its relation to the state in running the "Red" business makes the process of what Williams calls "mediation" profoundly nuanced. Investigation into this mediation is also meant to shed some light on the unsettled "Chinese characteristics" part of China's story of communication.

Commodification: is Marxism applicable?

Siding with Lefebvre and Williams, however, will prompt another theoretical and also a practical question: can Marxism be used in studying contemporary China anyway? This question must be answered before I approach "commodification," the most remarkable keyword in the book's title. This is not because commodification is a Marxist term, but because a considerable proportion of this work was informed by Marxism, specifically, Marx's critique of political economy. Before answering it, I want to make a brief detour, talking about what I mean, in an earlier statement, by "the world's second largest economy."

First and foremost, it means a great economic achievement of the Chinese state. It is common sense that ordinary Chinese people would benefit. According to a report of the World Bank, "[China] has lifted more than 500 million people out of poverty" (The World Bank, 2016). Echoing the world's second largest economy, China's tourism is also becoming the world's second largest travel and tourism economy after the United States (World Watch Institute, n.d.), and unsurprisingly, Chinese travelers turned out to be the world's biggest spenders (Cripps, 2013). Secondly, it leads to a long list of varied and wide-ranging challenges and issues such as social inequality, rural–urban disparity, the wealth gap, corruption, environmental crisis, rapid urbanization, etc. Consequently, these problems have resulted in a serious threat to the nation's socialist and traditional values, beliefs, ideologies, and political institutions of which Red Tourism is a part. Whether a "socialism with Chinese characteristics" or a "capitalism with Chinese characteristics" (Huang, 2008; Karmel, 1994), however, China does share many of those problems with contemporary Western countries where capitalism is dominant. Now I return to the question about the applicability of Marxism to China. My answer is in the affirmative.

Marxism is more of a critique of capitalism than a particular ideology (Eagleton, 2011; Harvey, 2010a). As such, Marx develops a well-grounded analytical framework to examine the capitalist market economy. It follows, then, that as long as capitalism is on the spot, Marxism works for any country, and no exception for China. This is why contemporary Marxist theorists such as David Harvey (2010a; 2010b; 2014) and Terry Eagleton (2011) continue making references to China when writing about capitalism. This is to say, although China is not a capitalist country in many regards, it would be wrong to discount the capitalistic forces related to the marketplace in analyzing Chinese mediascape. For example, "size" and "profit-orientation" certainly apply, even if "ownership" of media organizations does not, according to Herman and Chomsky's (1998/2008) propaganda model. Note that capitalism can refer to an economic system, a political system, or a combination of both. For the purpose of the current study, here I refer to capitalism as a mode of production in which privately owned capital determines the means of production. Commodification is an inevitable process in this mode.

Marx's whole analysis of capitalism starts from the concept of the commodity (Harvey, 2010a). It is an embodiment of two most critical elements of Marx's conceptual framework: class and labor on the one hand and a signifier of the capitalist mode of production on the other. The commodity is *the* fetish of capitalism. For Marx (1936), the term "fetish" simultaneously signifies two things: inherent magical powers of an inanimate object on the one hand and worship towards that object by people on the other. Marx (1936) uses the term "commodity fetishism" to address "the enigmatical character of the product of labor" (p. 82). Marx further explains the mysteries of commodities in that, "the social character of men's labor appears to them as an objective character stamped upon the product of that labor" (p. 83). It transforms the subjective, abstract aspects of economic value into objective, real things that people believe have intrinsic value. Deviating from Marx's notion of commodification in which Marx foregrounds labor process and surplus value, commodification in this study centers on the interplay between state power and capitalist power. And, the state and the capitalist have specific roles in Red Tourism.

One of the arguments of the book is that the state and the capitalist each has a Janus-faced role in producing propaganda. As a political entity, the state holds and exercises enormous political and institutional power, yet on the one hand, it is effectively the biggest investor. Harvey (2010b) articulates this point in the case of the state investing in infrastructures to maximize tax revenues. This is the way Red Tourism has snowballed and integrated from scattered small businesses into an industry. To put it another way, had the CPC not used huge chunks of tax money to undertake massive infrastructural projects such as high-speed railways, airports, and highways connecting Chinese metros to those Red Tourism destinations, Red Tourism would not have evolved from, say, a small heritage tourism business to a sunrise industry. Likewise, the capitalist also assumes dual roles in this study: a protagonist of capitalism on his own and a "surrogate" for the state, to borrow a term from Harvey (2010b). It has been the conventional wisdom that propaganda is a matter of state. But this is only partly true and this book aims to turn this wisdom on its head.

In Red Tourism, the capitalist fills the state's shoes to propagate intended ideological messages in typical capitalist fashion in which maximizing profit is the ultimate goal. This is an important but clandestine role of the capitalist in commodifying Red Tourism. The interwoven, intimate relationship between the state and the capitalist largely defines the oxymoronic nature of Red Tourism mentioned early on.

This book concludes with a political economic analysis of Red Tourism. Political economy can never be separated from the social space of Red Tourism. Therefore, the commodification of Red Tourism merits a separate chapter. It is typical to grasp the concept of propaganda as a purely political matter characterized by the predominant relationship between the state, government, and the public. In that case, propaganda is thought to be largely determined by the nature of a state's political system. This is why Siebert et al. (1963) normatively put China's media system into the category of "authoritarianism" in their famous book *Four Theories of the Press*. However, the current study argues that propaganda is not merely a machinery that serves "authoritarianism." It can be a commodity, an industry, and even an economy. And it can be very profitable. By virtue of that, it has enticed lots of entrepreneurs to join the force of the production of propaganda. Ironically, commercialism – the much-ballyhooed champion of capitalism once considered the arch enemy of communism in the Maoist era – has become an incubator for growing a new generation of propaganda. Rarely has the nexus between capitalism and propaganda been recognized, let alone a thorough political economic analysis of propaganda. In China, as I will try to show in the book, propaganda has been deeply connected to the economy ever since the beginning of socialist China, and now capitalism is making China's propaganda even more complicated in a way that has rendered our commonest knowledge about propaganda archaic.

Let me give an example. In January 2016, a giant golden-colored Mao statue in China was demolished by the government overnight a few days right before completion (Tatlow, 2016). Erected in the bare fields of a small village of Henan Province, the 120-foot-tall statue was built with money from Sun Qingxin, an entrepreneur, who spent about 3 million yuan ($465,000) on this project. The perplexity is twofold. First, generally presumed to be the production of the state, the propagandistic statue was created in the private sector instead. In other words, it would have made more sense if it was the other way around: the government built it first, then it was torn down by the villagers because of the land issue or other issues. Secondly, as the head of a manufacturing entrepreneur, Sun should pay for a statue of Deng Xiaoping, not Mao, because what Sun benefited from was market capitalism, not Mao's socialism. These seeming anomalies of propaganda confront Chinese people in everyday life, and it is this complexity that requires our close examination of propaganda beyond an ideological determinism.

A note on methodology

This study adopts a qualitatively interpretive approach, and two main source bases inform it. First, I drew from my own decade-long fieldwork in multiple locations in China, an extensive ethnographic data collection containing interviews, observation notes, photos, and collected documents. Besides formal interviews, either semi-structured interviews with the governmental officials and entrepreneurs, or unstructured interviews with the tourists on different tourist sites, the informants from the Red Tourism industry provided me with precious inside stories on such topics as land expropriation and labor issues. It should also be pointed out that the photos taken during my multiple field trips play a vital role in my analysis. This is, in part, due to the visual nature of tourism. That is, whether conceptualized as "gazing," "seeing," or "sighting," tourism is bound to the visual and particularly consumed visually (Urry, 1995, 2002). The image is powerfully communicative, so much so that it reminds us how Michel de Certeau (1984) read the everyday life of New York City from the top of the World Trade Center through *seeing*. McCarthy et al. (2015) raise a similar point in examining the visual field of the Barbadian postcolonial site. Secondly, I analyzed a wide range of popular discourse on Red Tourism culled from both traditional Chinese news media and the Internet.

Traveling widely across the country from 2011 to 2021 for this study, I eventually focused on one city, Yan'an, for this book. Located in the northwestern province of Shaanxi, Yan'an is better known as the "sacred place of the Chinese revolution" (*geming shengdi*). Yan'an was chosen because it holds the most prominent place in both the Chinese Revolution and Red Tourism. To a great extent, Yan'an is equivalent to Gettysburg in the American Civil War and Saratoga in the American Revolution insofar as it signifies the turning point of the war.

Partnered with the academic approach is my personal experience. This is not just relevant but valuable to the current study. No matter how conceptualized, China's propaganda is a cumulative space. "Cumulative" means requiring a long-time observation, long enough to cut across several periods. Born in the early 1970s, I experienced the last days of the Mao era along with the eras of Deng, Jiang, Hu, and Xi, together the five generations of Chinese leadership. Growing up in a propaganda ecology, propaganda to me was very much like the Internet to today's generation. I collected dung for the people's commune, attended yearly spring trips to the martyrs' cemetery in my school years, immersed myself in the sea of propaganda films, and later worked as a journalist for 16 years in a party's press organ. This long-term exposure and involvement in propaganda, however, did not make me a "puppet" of so-called "Communist propaganda." Contrarily, it has helped me foster a critical view on propaganda. Being critical here means not readily accepting opinions from academic authorities.

Nowadays, propaganda has become a sensitive topic in China. But it does not by virtue of what the Chinese people think what propaganda does. It does so by virtue of what the people who run things tend to cater to contemporary Westerners' thought on propaganda. It indicates an emergent hegemonic power of the Western model of propaganda. It also means that conducting this kind of research requires good strategies and techniques: for example, how to verify whether what the interviewee says is a true reflection or only meant to be politically correct, and how to sift through information in a way to separate cliché, platitudes, and rhetoric from real meanings behind. Here, I believe my age and journalistic experience help.

It is noteworthy that propaganda is the focus of the present study, not the effects of propaganda. Neither am I an advocate of the strong-effect theory, nor of the limited effects theory. The proposition that propaganda is a social space rejects any linear media effects model. The point that I wanted to stress, however, is that as a social space, propaganda is much more than an anti-democratic social practice on which every citizen may want to spit; neither does it automatically follow that any propaganda emanating from the state is deceptively evil, nor effective. The spatial perspective adopted in this book indicates a refusal of a conventional thesis of "forms of propaganda," putting propaganda into three clear-cut categories, namely, "white propaganda," "black propaganda," and "gray propaganda."

The reader may well be puzzled as to why a study ostensibly concerned with propaganda has devoted so little attention to propaganda tactics and techniques. True, I say very little in this book about these things. I will not focus on step-by-step "propaganda analysis" described in traditional propaganda research such as "10-step plan of propaganda analysis" in Jowett and O'Donnell's work (2012). This is because the objective of the current study is not to identify propaganda but to 3D print, for lack of a better word, among others, the social space of Red Tourism. Also, this is because that dimension has been already documented by a large number of empirical studies. Besides, techniques, strategies, and tactics in mass communication are innumerable and fluid. It is immensely challenging, and perhaps less meaningful for scholars to catch up on propaganda strategies no matter what cutting-edge propaganda analysis is in vogue.

Conclusion

Back to the metaphor of the tourist map that I mentioned earlier. The state and the public as sites are now connected by the routes of propaganda through which Red tourists travel between the revolutionary past to the post-socialist present. During their visits, their imaginations and memories about the state have been continuously reworked to make sense of the world. Dripping with all sorts of exhibition, performance, and interactivity, such reconfigurations of worldview and ideology taking place in the form of tourism seem no longer tedious. Conversely, Red Tourism is engaging and comforting regardless of its ideological nature.

This book treats Red Tourism as a media system somewhat like the news media. As such, the current study is not one of those some might call "tourism/ tourist studies" to the extent that it is neither a study of leisure, nor of the so-called "leisure class"; the book focuses on the system. Overall, this research is intended to be one of media studies. But it is a weird one, because it does not research into the system of the news media and censorship, but tourism. Moreover, seeing this media system as a social space and scrutinizing it with conceptual and analytical tools borrowed from cultural studies *and* political economy of media render this project even much weirder.

But this is necessary. For one thing, conventional models of propaganda apply poorly in China's case where propaganda is culturally deemed neutral. China's propaganda was not an outgrowth of the Communist ideology, in other words. It had prevailed throughout the imperial, republican, and socialist China. Thus, an ideology-predetermined approach to China's propaganda can yield nothing new except for a doomed propaganda state. For another, in contrast to the notion of social space, the habit of thinking propaganda as certain fixed media forms (e.g., newspaper, film, poster, television program, radio broadcast, etc.) and content elides the complexity of propaganda in which the relationships between the state and the public are constantly being twisted, reconfigured, and (re)produced by the state and also by the magic power of capital.

References

Anderson, B. (2006). *Imagined communities: Reflections on the origin and spread of nationalism*. Verso Books.

Babe, R. E. (2009). *Cultural studies and political economy: Toward a new integration.* Lexington Books.

Baudrillard, J. (1994). *Simulacra and simulation*. University of Michigan Press.

Bernays, E. L. (1928). *Propaganda*. Liveright.

Bernays, E. L. (1942). The marketing of national policies: A study of war propaganda. *The Journal of Marketing, 6*(3), 236–244.

Chao, R. (2013, October 9). Yan'an shengdi hegu: zhongguo hongse lvyou xin dibiao [Yan'an sacred land valley: A new landmark of Red Tourism in China]. *Zhongguo lvyou bao*, p. 16.

Cripps, K. (2013, April 12). Chinese travelers the world's biggest spenders. Retrieved from http://www.cnn.com/2013/04/05/travel/china-tourists-spend/.

Cunningham, S. B. (1992). Sorting out the ethics of propaganda. *Communication Studies, 43*(4), 233–245.

de Certeau, M. (1984). *The practice of everyday life*. University of California Press.

Denzin, N. K. (2008). *Searching for Yellowstone: Race, gender, family and memory in the postmodern west*. Left Coast Press.

Eagleton, T. (2011). *Why Marx was right*. Yale University Press.

Eco, U. (1997). Function and sign: The semiotics of architecture. In N. Leach (Ed.), *Rethinking architecture: A reader in cultural theory* (pp. 182–200). Routledge.

Ellul, J. (1965). *Propaganda: The formation of men's attitudes*. Knopf.

General Office of the CPC Central Committee and the State Council. (2011). National Red Tourism development outline 2011–2015. http://www.lotour.com/news/20110630/619116.shtml

Habermas, J. (1989). *The structural transformation of the public sphere: An inquiry into a category of bourgeois society*. MIT Press.

Hannigan, J. (1998). *Fantasy city: Pleasure and profit in the postmodern metropolis*. Routledge.

Harvey, D. (2003). *Paris, capital of modernity*. Routledge.

Harvey, D. (2010a). *A companion to Marx's capital*. Verso.

Harvey, D. (2010b). *The enigma of capital*. Oxford University Press.

Harvey, D. (2014). *Seventeen contradictions and the end of capitalism*. Profile Books.

Herman, E. S., & Chomsky, N. (1988/2008). *Manufacturing consent: The political economy of the mass media*. Pantheon.

Huang, Y. (2008). *Capitalism with Chinese characteristics: Entrepreneurship and the state*. Cambridge University Press.

Jin, Z. (2012, November 7). What makes Red Tourism so popular? The *People's Daily* (oversea edition). http://english.peopledaily.com.cn/90782/8009039.html

Jowett, G. S., & O'Donnell, V. J. (2012). *Propaganda and persuasion*. Sage.

Karmel, S. M. (1994). Emerging securities markets in China: Capitalism with Chinese characteristics. *The China Quarterly, 140*, 1105–1120.

Kim, K., & Barnett, G. A. (1996). The determinants of international news flow: A network analysis. *Communication Research, 23*(3), 323–352.

Koreneva, M. (2015, September 13). Russia draws in hordes of Chinese with "red tourism". Retrieved from http://news.yahoo.com/russia-draws-hordes-chinese-red-tourism-111346740--finance.html.

Kracauer, S. (1975). The mass ornament. *New German Critique, 5*, 67–76. https://doi.org/10.2307/487920

Lapiere, R. T., & Farnsworth, P. R. (1936). *Social psychology*. McGraw-Hill.

Lasswell, H. D. (1948). Propaganda. In E. Seligman (Ed.), *Encyclopedia of the social sciences* (Vol. 11). MacMillan.

Lefebvre, H. (1991). *The production of space*. Blackwell.

Liang, Q. (2010). *Qingdai xueshu gailun* 清代学术概论 [Introduction to the academic of the Qing dynasty]. Guangxi shifan daxue chubanshe.

Lin, C. (2015). Red Tourism: Rethinking propaganda as a social space. *Communication and Critical/Cultural Studies, 12*(3), 328–346.

Lin, C., & Nerone, J. (2015). The "great uncle of dissemination": Wilbur Schramm and communication study in China. In P. Simonson & D. W. Park (Eds.), *The international history of communication study* (pp. 396–415). Routledge.

MacCannell, D. (1999). *The tourist: A new theory of the leisure class*. University of California Press.

Marx, K. (1936). *Capital*. Random House.

Mattelart, A. (1996). *The invention of communication*. University of Minnesota Press.

Mattelart, A. (2000). *Networking the world, 1794–2000*. University of Minnesota Press.

Mattelart, A. (2008). Communications/excommunications: An interview with Armand Mattelart. *Review of International Studies, 34*, 21–42.

McCarthy, C., Greenhalgh-Spencer, H., Goel, K., Lin, C., Castro, M., Sanya, B., & Bulut, E. (2015). The visual field of Barbadian elite schooling: Towards postcolonial social aesthetics. In J. Fahey, H. Prosser, & M. Shaw (Eds.), *In the realm of the senses: Social aesthetics and the sensory dynamics of privilege* (pp. 137–169). Springer.

Nerone, J. (2015). *The media and public life: A history*. Polite.

Siebert, F. S., Peterson, T., & Schramm, W. (1963). *Four theories of the press: The authoritarian, libertarian, social responsibility, and Soviet communist concepts of what the press should be and do*. University of Illinois Press.

Socolow, M. J. (2007). "News is a weapon": Domestic radio propaganda and broadcast journalism in America, 1939–1944. *American Journalism, 24*(3), 109–131.

Tatlow, D. K. (2016, January 8). Golden Mao statue in China, nearly finished, is brought down by criticism. Retrieved from http://www.nytimes.com/2016/01/09/world/asia/china-mao-statue-henan.html?_r=0.

The World Bank. (2016, September 14). Overview of China. Retrieved from http://www.worldbank.org/en/country/china/overview.

Urry, J. (1995). *Consuming places*. Routledge.

Urry, J. (2002). *The tourist gaze*. Sage.

Welch, D. (Ed.). (1983). *Nazi propaganda: The power and the limitations*. Croom Helm.

Williams, R. (1961/1975). *The long revolution*. Greenwood.

Williams, R. (1980/1997). *Problems in materialism and culture*. Verso.

Williams, R. (2001/2006). Base and superstructure in Marxist cultural theory. In M. G. Durham & D. M. Kellner (Eds.), *Media and cultural studies: Keyworks* (pp. 130–143). Blackwell.

2 The problem of propaganda

This book was written at a time of China becoming the world's second largest economy and completed in the middle of the COVID-19 pandemic, where ideologies of all sorts, at all levels, in all countries, collide. The subject of this book, Red Tourism alongside propaganda, is part of this change. In the West, propaganda has been branded as anti-democratic by nature in academia and the public at large. Horkheimer and Adorno (2002) considered propaganda "antihuman" (p. 212). A typical study in this direction may examine propaganda content, methods, effects, institutional apparatuses, systems, strategies, gimmicks alike to strengthen democracy and morale. It follows that a media system dominated by mass propaganda generally hinders economic growth, and conversely, that economic liberalization will somehow bring about political liberalization and freedom of press. Under this contemporary Western liberal mentality, much of the US's imagination of China was based on the assumption that rapid economic development would ultimately lead to democratization, which Mann (2007) regarded as the "mainstream view of China in America today" (p. 7). But this did not happen. China's response and measures to prevent the spread of COVID-19 and its stance on the Russia–Ukraine war at the time of writing indicate that China will continue following its own economic and political paths without liberating its media system (Stockmann, 2013; Zhao, 1998). Another puzzle – central to this study – is why has Red Tourism, a propaganda project by the Chinese state, increasingly gained popularity at a time when market capitalism, rather than prior forms of socialism, is prevailing?

Drawing upon Chinese thought and scholarship on *xuanchuan*/propaganda and informed by a comparison to the Western history of propaganda studies, I demonstrate a deep conceptual and perceptual gap between propaganda and *xuanchuan* and argue that the gap is cultural. It cannot be sealed by translating *xuanchuan* into publicity, an increasingly popular treatment nowadays by government agencies and scholars. In addition, I will delve into what I refer to as *xuanchuan* culture. It is on the basis of this cultural practice that tourism and propaganda connect to, and thus speak to each other in this Red Tourism setting.

This chapter begins with a historical account of the development of propaganda studies in the West. This is necessary because we need a base and a structure for later comparison with *xuanchuan*. In discussion of the historical

DOI: 10.4324/9781003231783-2

and cultural roots of propaganda, the chapter directs special attention to the formation of what I refer as the "hegemonic Western model of propaganda." This is not merely a retelling of the grand narrative of propaganda studies; rather, this is a mapping out of nuanced relationships between propaganda and its elements embodied in a series of debates and controversies that may not be well heard elsewhere.

Following that, the chapter ventures into the world of *xuanchuan*. I argue that *xuanchuan* is neither a modern discovery nor an invention of the Communist Party of China (CPC). *Xuanchuan* is a cultural practice. Chinese people do not see *xuanchuan* as a malevolent force or "poison" but, contrarily, as "seeder," and ideally benevolent. The *xuanchuan* practice helps explain some critical yet uniquely perplexing problems about China's propaganda both in theory and in practice that cannot possibly be explained by any Western models of propaganda. Among others, why have some Chinese scholars advocated creating a "science" of propaganda? Why do some Chinese communication scholars still hold that news and *xuanchuan*/propaganda are largely the same? Another intriguing question is, why modern Chinese journalists still proudly call themselves the "people doing *xuanchuan*/propaganda"? I also bring ideology, hegemony, and worldview into light. By examining their connections and disconnections in the realm of representation, I want to relocate propaganda from the realm of ideology, its place of origin, to a wider realm of popular culture where Red Tourism thrives.

This chapter concludes in an attempt to discuss the Chinese characteristic forms of governance and mass communication in relation to propaganda. Moving between and extending what I have discussed earlier in this chapter, I argue that China is a propaganda state. Note that what I mean by a "propaganda state" here is not the same thing described by Western news media/academia. My argument is based on a form of governance and a specific culture rooted in the long history of China. Let's start with the folklore of propaganda.

Propaganda as deceptive manipulations

Several communication scholars purported that we live in an age of propaganda (e.g., Lindeman & Miller, 1940; Pratkanis & Aronson, 2001). But what is propaganda? Bytwerk (1998) called it an "awkward question" (p. 158). This awkwardness stems from the abundant literature and thought on the topic, and despite that our collective inability reaches a common, satisfactory definition of the term (Jowett, 1987). This is in part because scholars often see propaganda as multifarious "things" such as technique (Lasswell, 1927; Miller, 1937), method (Bernays, 1928; Ellul, 1965), practice (Lapiere & Farnsworth, 1936), and ideology (Herman & Chomsky, 1988/2008), among others. But there is a common ground among scholars: propaganda is the deliberate manipulation of people's minds. Lasswell (1927) defined propaganda as the "management of collective attitudes by the manipulation of significant symbols" (p. 627). Clyde Miller (1937), the director of the Institute for Propaganda Analysis (IPA), told Americans that propaganda is meant to manipulate individuals' or groups' opinions or actions.

Lapiere and Farnsworth (1936) referred to propaganda as deliberate attempts to achieve desired effects by the propagandist.

The historical roots of propaganda

The miasmic aura of propaganda was not always inherent in Western cultures. Propaganda is a Latin term, initially referring to the production of plants and animals (Fellows, 1959; Hosterman, 1981). The word "propaganda" first appeared in English in 1718 with the same neutral connotation as it was in "de propaganda fide" (Fellows, 1959). Accordingly, the English term "propaganda" originally seemed rather "benevolent" in a religious sense. Gordon (1971) noted that propaganda was not used as a pejorative two generations ago. This is consistent with Mattelart's (1996) account of the origins of propaganda in the West – De Propaganda Fide of 1622 – as a practice of spreading God's words to the non-Catholic world. Starting from that point, the word referred to the practice of propagating certain doctrines and was considered a "perfectly legitimate form of human activity" to disseminate the truth (Bernays, 1928, p. 22). By the 19th century, the word "propaganda" still held its neutrality (Jowett, 1987).

The re-signification of propaganda as large-scale political propaganda was believed to have developed around the time of the French Revolution in the late 18th century (Dowd, 1951; Rogers, 1949). However, at least one author suggested, though not implicitly, that the starting point of modern mass propaganda in the West might be earlier than that. Sawyer (1991) remarked that political pamphlets for propaganda purposes, which he calls "printed poison," were massively produced and distributed in 17th century France.

Communication scholars, however, have arrived at a fairly unanimous conclusion that the First World War was the genesis of modern mass propaganda (e.g., Fellows, 1957, 1959; Jowett, 1987; Sproule, 1987; Whitton, 1951). Creel (1941) regarded propaganda as a "whipping boy" for manufactured hate of the Great War (p. 341). Lasswell (1927) called it the "discovery" of WWI. This discovery, as Edward Bernays (1942), the father of public relations, explained, was the realization by the belligerent powers that ideas are just as powerful as arms.

In light of this discovery, persuasion industries emerged as marketing exploitation of people's fear and desire in the wake of the First World War. As a result, propaganda transfigured into different things in different spheres: advertising in the commercial sphere and citizenship education in the educational sphere (Gordon, 1971). Advertising companies from the private sector mushroomed hand-in-hand with propaganda institutions established by the state. Institutional advertising used propaganda techniques to create a favorable image of the company (Pearlin & Rosenberg, 1952). George Creel's book *How we advertised America* highlighted the intimacy between propaganda and advertising. Creel was the head of the Committee on Public Information (CPI). Created by President Woodrow Wilson during World War I, CPI was a propaganda organization whose official propaganda campaign during the postwar period is believed to be the "immediate impetus" for American re-interpretation of the dark side of propaganda (Sproule,

1987, p. 63). Creel (1972) proudly described the role of the CPI as a "plain publicity proposition, a vast enterprise in salesmanship, the world's greatest adventure in advertising" (p. 4). In Lasswell's (1927) view, even propaganda itself had become an industry with the emergence of an army of professionals including propaganda practitioners, professors, and teachers. Unsurprisingly, the American news industry was also swayed by this industrialized propaganda tide. Dewey (1918) claimed that "[o]ne almost wonders whether the word 'news' is not destined to be replaced by the word propaganda" (p. 216). This is particularly true for US radio journalism of the Second World War. Socolow (2007) has remarked that the head of propaganda radio held that "the journalistic function of American propaganda and the propagandistic function of American journalism were inseparable" (p. 126).

The hegemonic Western model of propaganda

The Western model of propaganda is the end product of fears. It emerged as an intellectual response to growing fears over war propaganda of World War I and rapidly developed after WWII. The model crystalized as fears suffused globally during the Cold War period. It was (re)modeled in different ways by different scholars, but all versions bore deep fears. They were fears over Western democracy being victimized by propaganda. The Western model of propaganda was exported to the developing world along with the Western ideas of journalism, public relations, and mass communication. Fears were also packed in that delivered package. As a cumulative effect, propaganda had been universalized systematically and internationally into a firm rubric for legitimately despising journalisms, news practices, and media systems of non-Western others. Like hegemony of any kind, the Western model of propaganda wasn't made from whole cloth. Propaganda analysis was its hotbed, and the model was constructed based on propaganda studies.

Initially, propaganda analysis set the tone for the development of the Western model of propaganda. As a paradigm, propaganda analysis germinated during the post-World War I period. Sproule (1987) pointed out that it was a critical paradigm created under American progressivism in an effort to measure mass media effects brought about to the modernization of American society. Early American progressive thinkers such as Will Irwin, Walter Lippmann, and George Seldes resisted the neutral thesis of propaganda proposed by American "propagandists" such as Lasswell and Bernays (Sproule, 1989a, 1997). Under this influence, Americans considered propaganda the "corrosive product," "austerely and pharisaically demanding its exorcism from American life" (Creel, 1941, p. 341). This sounds like a political movement. As a matter of fact, Garber (1942) did call it the "propaganda-analysis movement," arguing that propaganda analysis paid much attention to verbal tricks without linking the problems to a larger social context in which they present (p. 241). The reproachful tone toward propaganda was hardened as propaganda studies progressed. This is because, in part, propaganda studies, too, stemmed from fears. Those were fears over the malevolent use of mass media and its effects during two World Wars. Fears had profound effects on the nature of the study of propaganda as well as the formation of the Western model of propaganda.

First, as an immediate effect psychology became the dominant paradigm for studying propaganda. During the post WWII period, propaganda theory was dramatically shaped by the study of mass psychology, largely by conditioning theory and psychoanalysis (Garber, 1942). Accordingly, vocabularies of propaganda studies were unsurprisingly psychological: "manipulation," "collective attitudes," "deliberation," and "persuasion" to name a few. This trend is manifested in Jowett and O'Donnell's (2012) definition of propaganda that "propaganda is the deliberate and systematic attempt to shape perceptions, manipulate cognitions, and direct behavior to achieve a response that furthers the desired intent of the propagandist" (p. 4). It sounds as if propaganda were an area of study in psychology. Lapiere and Farnsworth (1936) likewise suggested that definitions of the term "propaganda," however different, "must be psychological" (p. 71). Moreover, the notion of psychological warfare rendered propaganda studies extremely psychological (Doob, 1935; Linebarger, 1948; McLaurin, 1982; Qualter, 1962; Szunyogh, 1955).

Secondly, in the long run, the academic trend for propaganda studies rose and fell correspondingly somewhat dependent on the fluctuation of fears. In light of fears of Nazi propaganda, propaganda studies emerged and the paradigm of propaganda analysis thrived from 1919 to 1937 (Sproule, 1987). Debates on the neutrality of the term appeared during the post-World War II period when fears subsided. With the Communist bloc gaining power, fears increased again, consequently generating more propaganda institutions and studies. And the study of propaganda itself turned out to be propagandistic. Parry-Giles (1994), for example, has contended that American propaganda studies secured its prominence during the Cold War because it tested "how well the United States was faring in the war of words" (p. 204). Chinese communication scholar Liu Hailong (2013) argued that propaganda studies itself transformed into propaganda during the Cold War because much of such academic effort was bound to add fuel to the hostilities. The subfield of propaganda study in communication had become less promising upon the collapse of the Eastern Bloc when fears over communism were allayed.

Related is the opportunistic nature of propaganda research. This is about taking advantage of situations in order to get funding from whatever source. Propaganda studies flourished in the West only when government had poured huge chunks of money into propaganda institutions such as CPI, IPA, the United States Information Agency (USIA), and the like. Propaganda studies as a field died in want of funding and the topic by itself became obsolete in communication studies.

Fourth, fears and government funding together contributed to the ideology-oriented-and-driven nature of propaganda research. Jowett (1987) identified three challenges in the study of propaganda: (1) defining propaganda, (2) erasing the negative connotation of propaganda, and (3) creating a body of systematic literature on the topic. All of these problems can be traced back to ideology. Consequently, propaganda studies had turned into a sort of black hole of the field: on the one hand, it contributed to the classical theories in the field (e.g., strong, weak/limited media effects) and made the choice of the dominant paradigm for the field (e.g., the social scientific, empirical); on the other hand, nowadays propaganda research in the field of communication has become increasingly invisible. This "black hole" can be interpreted as, to borrow from Sproule (1989a, p. 1), a "long disparaged but quite rich legacy." In a nutshell, propaganda is not a safe sphere of inquiry (Jowett, 1987). Perhaps a safer way to deal with propaganda is to treat it as weaponry. Then the study of propaganda, in a sense, turned into something like a weapon test site: propaganda material and data were collected, assessed, and analyzed in order to gain the strategic edge over hostile countries. As a result, many propaganda studies have focused on examining various propaganda strategies, tactics, techniques, and practices (Kenez, 1985). In a very real sense, propaganda studies was born and obscured in the war of ideology.

There was an exception. Propaganda was rethought as something neutral by few scholars in the United States somewhere between the 1930s and the 1940s (e.g., Fellows, 1957; Lapiere & Farnsworth, 1936; Sproule, 1989b). It reflects American social science's struggles between behaviorism and a value-free approach (Black, 2001). During this period, propaganda practitioners and scholars such as Bernays (1942), LaPiere and Farnsworth (1936), and Lasswell (1948), among others, called for a neutral definition of propaganda. Some went even further, arguing that propaganda even can work positively for Western democracy. Bernays (1928), for example, claimed that careful and clever manipulation of public opinions is essential to democratic society. Similarly, Perry (1942) reckoned that propaganda is necessary for a democratic government to function effectively.

Nevertheless, the neutrality counterthesis of propaganda had hardly affected the popular imagination of propaganda as "poison." For one thing, although the voice of the neutrality advocates could be heard, the examples and cases the scholars used to make their points were basically either political propaganda or war propaganda; those by no means were neutral. For another, there was a tendency in the neutral thesis, explicit or implicit, to indicate that propaganda needs to be avoided. Take Ellul's work for instance. Ellul's (1965) book *Propaganda: The formation of men's attitudes* has to be critical to propaganda studies in many aspects. For Ellul, propaganda is a sociological phenomenon; propaganda exists in any society. In democratic societies, something that Ellul called "democratic propaganda" prevails. He used words like "channel," "shape," and "adjust" to refer such type of propaganda, avoiding using "manipulate." More startlingly, Ellul (1965) claimed that "propaganda, regardless of origin, destroys man's personality and freedom" (p. 137). To be fair, in this book Ellul conceptualized various propagandas and kept reminding us of various purposes of propaganda, good or bad. Rarely, can any attentive reader recall an example of "good" propaganda. The neutrality thesis still can be found in scholarly work later from time to time (e.g., Black, 2001; Jowett & O'Donnell, 2012; Rohatyn, 1988). Jowett and O'Donnell (2012) have made it clearly that "[p]ropaganda is not necessarily an evil thing. It can only be evaluated within its own context according to the players, the played upon, and its purpose" (p. 367). When it comes to the American Revolution, however, the authors do not tend to use the word "propaganda," preferring a long noun phrase, "the spread of ideas through the printed word" (Jowett & O'Donnell, 2012, p. 79). This common strategy of differing propaganda from other callings in the neutrality in different contexts has rendered the connotation of propaganda hopelessly negative even in a natural thesis. In addition, it is hard to find any empirical study that really puts propaganda into a neutral category. Then, what is the Western model of propaganda, anyway?

The Western model of propaganda derives from war propaganda analysis and studies throughout two World Wars in the early 20th century. It was generated from fears, generated much fear, and ultimately was consolidated in fears. Under the ideology of American progressivism, this model postulates that propaganda is the deliberate manipulation of the public opinion and thus fatal to both Western liberalism and democracy. Upon its formation, the Western model became the "dominant rubric" for mass communication research (Sproule, 1989a, p. 1). This model of propaganda was developed within the social science paradigm and fixed in the context of psychological warfare. But social sciences and psychology did not rescue propaganda studies from obscurity. This is, to a large extent, because this model is super-ideological. Social scientists and psychologists had opted for a neutral world of persuasion to avoid being contaminated by the value-and-ideology-laden paradigm. This model describes propaganda outside Western democracy as much more nefarious than that within it. It has rendered propaganda studies deterministic and thusly less meaningful. Cumulatively, propaganda studies became a ghost of the field.

There is another model of propaganda from political economy of communication. Political economy has contributed enormously to our understanding of propaganda and the power relations in producing, distributing, and circulating propaganda. Not only has this scholarship enriched traditional propaganda studies with non-discourse (political economic) analysis, it has also considerably widened the scope of the study of propaganda from microscopic textuality to macroscopic systems-orientation. Herman and Chomsky's propaganda model is a milestone in that regard. The propaganda model suggests that content of US mainstream news outlets become propaganda through five filters: (1) media ownership, (2) advertising dependency, (3) heavy reliance on government news sources, (4) flak, and (5) dominant ideology (Herman & Chomsky, 1998/2008). In their modeling, the five filters are really two filters: capitalism and ideology. This is a particularly illuminating way to rethink propaganda not as a "thing" but as a space, produced by inequality of wealth and power and producing the same things.

Nevertheless, it is quite a different story in China. In what follows, I will examine *xuanchuan*, the parallel of Western propaganda. Highlighting the fundamental differences in major dimensions between *xuanchuan* and the Western notion of propaganda, I propose an alternative model of propaganda, the "*xuanchuan* model." This model is intended to demystify China's propaganda that has been fantasized as an outgrowth of the Communist ideology. I argue that *xuanchuan* in China is a cultural tradition, not a modern invention. The idea of *xuanchuan* complicates the oversimplified hostile relations in the West between propaganda and other crucial elements of an open society.

Xuanchuan as a cultural practice

The hegemonic power of the Western model of propaganda has reached China. The Propaganda Department of the CPC changed its English name to the "Publicity Department" though its Chinese name remains unchanged. It is ironic that by switching the translation from propaganda to publicity, the government bureau covertly accepted the Western hegemonic perspective of propaganda, which sees the bureau's work as deceptive, undemocratic. Chinese communication scholar Liu Hailong (2013) has gone further, calling what he refers to as the "US model" the "scientific propaganda model" (p. 5). For Liu, characterized by the systems of government spokesperson and public relations, the "US model" is "scientific" because it prioritizes a neutral way to deal with propaganda, which is called "publicity." Liu is not alone. It has been a common strategy for scholars to translate "*xuanchuan*" into "publicity" when preferring a neutral connotation, and "propaganda" for the opposite. Such treatment, though practical, has created more problems than it solved. The opportunity to rethink China's propaganda in relation to its economy, culture, and society has been lost in such translation that embraces an old, ahistoric, and acultural mode of thinking.

In this section, I argue that *xuanchuan* is a cultural practice. As such, propaganda in China has its own origins, tradition, history, practice, characteristics, debates, and concerns that do not necessarily fit in the Western model. Theories of propaganda developed in the West are insufficient for scrutinizing a society where propaganda as a cultural practice is believed to infiltrate and affect every stratum.

In what follows, I start with a brief historical account of *xuanchuan*. It is meant to challenge the hegemony of the Western model of propaganda on the one hand and to outline the cultural practice of *xuanchuan* by focusing on its four characteristics on the other. I then dive into what this cultural practice means to what some Chinese scholars refer to as the "Chinese characteristics of communication," with emphasis on the long-lasting debate on the relation between *xuanchuan* and journalism. This is necessary since it builds the foundation on which I will wed tourism with propaganda and journalism in the next chapter, arguing that tourism can be better understood as a propaganda system, akin to the news media. I will also explain why there has been such a *xuanchuan* culture in China that did not appear in the West. Finally, I will end this section by bringing to light three big confusing but closely related concepts: ideology, hegemony, and worldview. I conclude this section with sketching out the *xuanchuan* model, a rather tentative solution to the problem of propaganda.

The cultural roots of xuanchuan

Xuanchuan is the Chinese word for propaganda. Unlike its Western parallel, *xuanchuan* is a neutral term without any derogatory connotation (Lin, 2015; Lin & Nerone, 2015). The term is composed of two Chinese characters, *xuan* 宣 and *chuan* 传. *Xuan* means "announce" and *chuan* "disseminate." According to the *Modern Chinese dictionary*, *xuanchuan* carries three meanings: to announce, to explain or educate, and to propagandize. In practice it commonly refers to three things: propaganda, advertising, and public relations. Communication scholar Robert McChesney referred to these as "insincere communication" for their manipulative nature. Nevertheless, the Chinese people call these things "*xuanchaun*/propaganda" because all of them *do not just communicate* (as exchanging information) but *disseminate* information for specific purposes. Originally, the word "*xuan*" in Chinese ancient writings means to spread information from rulers to the ruled (Cao, 1987). Later this meaning was associated with conveying orders from the top authority downward (Liu, 2013). The authoritative power has always been assumed by those who do *xuanchuan*. In other words, *xuanchuan* signifies a kind of power merged from an authoritarian culture. To the extent that *xuanchuan* implies concentrated power, it does share a marked characteristic with the modern Western notion of propaganda.

But x*uanchuan* has different cultural roots. The origins of *xuanchuan* as a social practice started much earlier than its Western counterpart. The World Wars' timeline of the development of propaganda studies in the West is irrelevant to the history of *xuanchuan*. While scholars did remark that propaganda activities in the West also have a longer history, they have concluded that propaganda as a means of mass persuasion/manipulation is modern (e.g., Block, 1948; Qualter, 1962; Schettler, 1950). Regarding propaganda as a regular department of government, for example, Whitton (1951) referred to it as the "great innovation of modern times" (p. 142). Likewise, Ellul (1965) held that the new mass media is the precondition for the coming of modern propaganda. But many Chinese communication scholars disagree with this in many respects.

They believe *xuanchuan*/propaganda, in the full sense of the word, is ancient (Dai, 1992; Li & Guo, 1992; Qiu, 1993; Shi & Gao, 2011). The creation of *xuanchuan* as a practice appeared in the Spring and Autumn Period (770–476 B.C.E.). This was not persuasion by ad-hoc communication or word of mouth; it involved mass persuasion, mass movement, and mass transportation. The massive, systematic, and political focus of *xuanchuan* during this period bears a striking resemblance to modern Western propaganda even though it predates modern mass media. Qiu (1993) has argued that the *xuanchuan* practice reached its zenith in ancient China, perhaps during the Spring and Autumn Period. Many factors are believed to contribute to the spectacular rise of *xuanchuan* at scale. Among many Chinese scholars, Liang Qichao's (1873–1929) account is illuminatingly evocative.

Liang (1923) suggested that the increasing mobility of the people paved the way for the coming of the *xuanchuan* practice. Liang made an interesting point that in the absence of mass media, *xuanchuan* was carried out by mass movement of social elites including officials, merchants, and intellectuals. This is to say, in an archetypical *xuanchuan*, the members of the elite class played the role of the medium in propagating information to the common folk by word of mouth, which Raymond Williams (2003) referred to as social communication. This tradition saw great success during the Chinese Revolution in the 20th century. Constrained by illiteracy and materials, the CPC propagandized revolutionary ideas and its policies to the proletarian class in a manner heavily dependent on oral communication and in-person visits. To explain what propaganda is, Mao (1991) said, "a person as long as he talks to others, he is doing propaganda work" (p. 838). Even during the Cultural Revolution, when all forms of mass media transformed into propaganda machinery, word-of-mouth propaganda was still preeminently effective. Chu (1977) has pointed out that the Chinese way of propaganda was not through the news media, but rather a case of "almost everyone talking to everyone else" (p. 4). Dymkov (1967) also observed that compared to the leading role of oral propaganda, printed propaganda only played a supporting role in the Cultural Revolution. The examples are multitude. The point is that the precondition for *xuanchuan* was not mass media as was true in the West; it was human mobility combined with word-of-mouth communication.[1]

The democratization of aristocratic schooling made the *xuanchuan* practice systematically massive. Liang Qichao (1923) noted that aristocratic schooling became increasingly scattered in the civil society during the Spring and Autumn Period, enabling the ideology of the aristocracy to reach out to a much wider audience. In addition, new writing material further boosted the *xuanchuan* practice (Guo, 1985). Unlike the previous bamboo and wooden slips, silk paper came into being, which was much easier to transport.

Meanwhile, the Chinese school of *xuanchuan* emerged. The major intellectuals of this school were generally pre-Qin thinkers including Confucius and Mencius, among many others. The *xuanchuan* school is best known for its humanism. Members of the school stressed that *xuanchuan* was the best means of governance. They believed education and propaganda must be given priority to govern a state. Mencius, for example, claimed that the Tao of governance is to gain the people's support that can only be achieved through education and propaganda (*Zhuzi jicheng*, 2006). This school of thought on *xuanchuan* has far-reaching sociopolitical consequences.

As an immediate impact, *xuanchuan* moved up to a massive scale and with great social penetration (Deng, 1988). Take *zhong* 忠 (loyalty) and *xiao* 孝 (filial piety) propaganda for instance. They are considered two types of social identification in Confucianism (Hwang, 1999). The *zhong-xiao* propaganda was regarded as the core value of the Ming and Qing dynasties and believed to be "valuable experience in running and uniting the state" (Gu, 2014, p. 23). It is still the case in contemporary China. In the state sector, being loyal to the country has been put into the "core socialist value system" under the presidency of Hu Jintao, whereas filial piety has been deemed as the key to achieving a harmonious society in the private sector. Together, *zhong* and *xiao* provide the ideological and moral foundations for post-Mao China. Like any culture, *xuanchuan* has a pattern that can be characterized by four characteristics: (1) social integration, (2) education, (3) human mobility, and (4) social networking.

Four characteristics of xuanchuan culture

The upshot of ancient *xuanchuan* activities was the emergence of an authoritarian culture, which I would call *xuanchuan* culture. It is a form of authoritarianism.

Nevertheless, it is not defined by an authoritarian political system, but by the authoritarian way of life. This is to say, *xuanchuan* is a culture, a lifestyle, and a form of mediation. Culture has countless meanings; herein I refer to the meaning of culture put forth by James Carey (1989), who regarded communication as culture for it is the "symbolic production of reality" (p. 23). A vivid example of authoritarian culture/communication occurs at most levels of Chinese societies when parents make choices and decisions about colleges, jobs, and even marriages on their children's behalf. As a cultural practice, *xuanchuan* prevailed and is still prevailing in all Chinese societies throughout imperial, republican, socialist, and post-socialist China.

However, this is not the place for a comprehensive overview of this culture. The *xuanchuan* culture in China is too big to be told without gross oversimplification, and my treatment makes no pretense of completeness. Here, I identify and highlight a series of characteristics of this culture, as they sketch out what the culture looks like. Four characteristics that define the *xuanchuan* culture are: (1) the function of social integration, (2) the education pathway, (3) the emphasis on human mobility, and (4) the conventional social networking method. I will elaborate these in the following.

At the heart of the *xuanchuan* practice is social integration. *Xuanchuan* seeks to integrate people into an intended society, for better or worse, under a certain set of ideology. It is the basis on which *xuanchuan* became a cultural axiom for governing and retaining social harmony throughout China's history. This is why, starting from the Kuomintang (KMT) regime, the Chinese government has self-consciously used the term *xuanchuan*/propaganda in describing its system and the department dealing with the subject. This idea of integration, however, was approached by different scholars from different angles, yet was rarely pinpointed and largely neglected. In examining the organizational structure of China's political propaganda system, Shambaugh (2007) noted that most Chinese citizens do not take any negative connotation from propaganda because "it believes should be transmitted to, and inculcated in, various sectors of the populace" (p. 29). But rather than viewing it as a cultural belief held by many, he mistakenly attributes this neutral attitude to the Party's view that propaganda is a legitimate tool to build the society. In his account, the masses seem to be "brainwashed," believing that propaganda is neutral. Arguably, Ellul's (1965) idea of integration propaganda is much closer to *xuanchuan*. For Ellul, integration propaganda aims at unifying social groups by imposing a pattern. This otherwise stimulating idea loses its luster when juxtaposed as an antithesis of agitation propaganda in a dichotomy, among many others (e.g., active/passive, direct/indirect, political/sociological, vertical/horizontal, rational/irrational propaganda, etc.). In Ellul's scheme, the relationship between agitation propaganda and integration propaganda is that the latter replaces the former. This leads Zhang and Cameron (2004) to argue that a structural transformation of Chinese propaganda from agitation to integration is underway. In my view agitation is a propaganda technique, whereas integration is a function of propaganda. They are not two types of propaganda, and therefore not mutually exclusive; rather, they are two dimensions of propaganda. Typically produced in the form of catchy slogans, agitation propaganda such as "Be All You Can Be," "Army of One," or "Army Strong" can be easily spotted or heard even in a democratic country like the United States. As a technique, agitation propaganda has never been replaced and outdated. It just moves from one space to another.

Another characteristic of *xuanchuan* is its inherent connection to education. But education in this context has specific meanings. First, education can be seen as what the Chinese people call "*jiao hua* 教化", a combination of two Chinese characters, "education" and "change", hence "education for change." Unlike its English translation, the Chinese term sounds old-fashioned without any modern rhetorical quality and mostly appears in polemics. Believed to be an amalgam of politics, morals, and education, *jiao hua* is sometimes loosely translated as "enlightenment." The second character "*hua* 化" (change) carries all the weight of its cultural signification. Historically, *jiao hua* refers to the change of minds to be submissive by conforming to certain norms and values, whether ethical, moral, cultural, social, or political. This is exactly what Weber (1964) believed Confucianism was about. Coincidently, Mao Zhedong shared a similar view with Weber on Confucianism for which Mao encouraged the public to criticize Confucius and Confucianism during the Cultural Revolution, resulting in the "Criticizing Lin and Confucius Campaign."

The second implication is that education, formal or informal, is believed to be the pathway to propagandize. This is why propaganda in China sometimes is referred to "political/ideological education." Under the same rationale the system implementing propaganda and the education system used to be united in a single system called "*xuanjiao xitong* 宣教系统" (propaganda and education system) in the Mao Era. Propaganda's role of education can be reached by Williams's idea of education. For Williams (1961/1975), education is a process of selection and distribution. What is to be selected or omitted for distribution is partly cultural and partly ideological. Williams pointed out that, while some consider this process education, others from different cultures may regard it as indoctrination. This exactly is the case in China. Williams (1961/1975) depicted such indoctrination as "the transmission of a particular system of values, in the field of group loyalty, authority, justice, and living purpose" (p. 126). This depiction is what Chinese people generally think about what propaganda does. In China, the educational signification of propaganda was passed down from generation to generation as a cultural practice of *xuanchuan* and continued evolving from imperial periods to modern times.

Western scholars have also acknowledged the educational dimension of propaganda. Gordon (1971) blurred the previously defined line between propaganda and education, referring to teachers as "propagandists" and textbooks as "instruments of propaganda." Gordon (1971) argued that juvenile delinquency in the United States was not a failure of education, but the breakdown of the nation's propaganda for not providing the "right kind of cultural indoctrination" (p. 167). He concluded that the big problem of the US education system was too little propaganda in early schooling and too much in higher education. The logic here is fairly simple: when kids are young, they need more propaganda to guide them to behave morally, legally, and responsibly, whereas when they get mature and are competent at reasoning, less propaganda is needed. Nevertheless, Gordon's view of the connection between propaganda and education was heterodox in the American context. The popular belief in the West has been just the opposite: propaganda aims to enslave, while education serves to empower the individual (Doob, 1935, 1949; Martin, 1929; Qualter, 1962).

Mass mobility is another feature of *xuanchuan*. It marks off *xuanchuan* from its Western parallel in both history and practice. In the West, propaganda is believed to be preconditioned by wide circulation and distribution of mass media. In other words, the proliferation of printing technology alongside the rise of literacy is where the modern Western history of propaganda began. But this is not the case for *xuanchuan*. The driving force behind the formation of *xuanchuan* was greater human mobility brought about by the extension of transportation networks (e.g., roads, passes, canals) in the wake of frequent interstate wars in ancient China (Lin, 2015; Qiu, 1993). Instead of circulation through media, people circulated and propagandized, acting as propaganda vehicles or media. In doing so, the literacy barrier was circumvented by word-of-mouth. In practice, the *xuanchuan* tradition of traveling, or mapping space by human travel, remained effective during the early period of the Chinese Revolution. The Communist propaganda deeply penetrated the countryside not by regular forms of mass media, but by tens of thousands of "propagandists" at all levels of the Party leadership constantly traveling. This mass traveling tradition of *xuanchuan* reached another peak during the Cultural Revolution. Communication-wise, the unprecedented element in what some call the "unprecedented Cultural Revolution" can be fairly defined by the unprecedented volume and speed of mass mobility embodied in a series of political yet human movements including the "Great Mass Rally," "Transfer of Cadres to Lower Levels," "Up to the Mountains, Down to the Villages," and so forth. But the story does not stop there. Mass mobilization of cadres for a propaganda effort still characterizes the CPC's propaganda work in post-Mao China. Under the system of "Propaganda Teaching Team" (*xuan jiang tuan* 宣讲团), tens of thousands of cadres nowadays are routinely traveling from cities to rural areas, engaging in diverse *xuanchuan* activities at the grassroots level. There is little exaggeration in saying that *xuanchuan* is propaganda in motion.

This leads to the social networking of *xuanchuan*. It suggests that people, the human agents, are of paramount importance in the cultural setting of *xuanchuan*, which renders mass media less glittering in terms of influence and effectiveness. Arguably, the great success of the CPC's propaganda system was not its news media system, but a parallel system one might call the cadre-based human social network. It resulted in an ever-increasing atomization of society: the top Party leadership assigned its propaganda tasks, passed some of its own responsibilities down to millions of Party cadres, and finally reached out to the masses. In a sense, the social networking method of *xuanchuan* is somewhat like the modern multilevel marketing (MLM) network, except the products were not goods or services, but highly uniformed indoctrinations and instructions. It was indeed a hierarchy of multiple levels of socially propagating networks characterized with the word-of-mouth approach of interpersonal communication. Ostensibly, the social networking method can be explained by the theory of personal influence, which, as advanced by Katz and Lazarsfeld, stresses communication from person to person. Central to that theory is a model called the "two-step flow of communication," suggesting that mass media first affect opinion leaders and then the leaders affect wider populaces (Katz & Lazarsfeld, 2006). Nevertheless, this kind of vertical influence is only one side of the social networking approach of *xuanchuan*. The other side that can better characterize *xuanchuan* is horizontal influence. It is not wielded by "opinion leaders," but, in many cases, by family members, relatives, and close friends. In contrast to vertical influence, the horizontal influence is more powerful and effective. This is why as a business model, MLM was banned in China in 1998 after being accused of undermining social stability and economic order by the Chinese government (Wong, 2002).

The triangle of dissemination

In the field of Chinese communication study, *xuanchuan*, journalism, and communication make up the triangle of dissemination. The *xuanchuan* culture along with the four characteristics created a different academic milieu for propaganda studies in China. In the West, tremendous effort has been made to studying propaganda effects for the purpose of avoiding propaganda, whereas in China, the dominant research paradigm has been the one aiming at improving propaganda work in practice. The cultural gap between propaganda and *xuanchuan* also manifests in a long-lasting and still ongoing academic debate about the relationship between propaganda and journalism. The central question of the debate is this: should journalism be considered propaganda? The seemingly non-issue from the Western perspective has generated a substantial body of literature. Using "*xinwen yu xuanchuan* 新闻与宣传" (journalism and propaganda) as a keyword to search titles of academic journal articles on China National Knowledge Infrastructure (CNKI) databases, 937 items were found. Why have those Chinese authors been so obsessed with such an ostensibly simple question? For most Westerners, journalism is certainly not propaganda.

But many Chinese scholars hold the opposite view. Whether arguing that journalism intersects propaganda, or that one includes the other, the consensus view is that propaganda and journalism are tied to each other. At the heart of this view is the conviction that ideally journalism is meant to educate the people to integrate them into a better society. By virtue of this, Chinese journalism scholars have emphasized the humanistic side of journalism in one way or another. For example, the major concern of the commercialization of journalism in China has been the "decrease of humanistic content," whereas the "humanism of journalism" is taken for granted (Tong, 2001). The Chinese humanistic view of journalism is radically different from the Western liberal idea of journalism, which can be understood as the representation of public opinion (Nerone, 2015). In other words, journalism in China has been, to a large extent, deemed as a form of guidance, not a form of representation. As such, China's journalism operates in line with the *xuanchuan* practice in many respects. In addition to their overlapped purposes of integration and education, journalism in China also shares an authoritative tone with propaganda in practice. It has been a tradition that Chinese journalists do not prioritize presenting balanced opinions, as is true in the Western journalistic practice; they only offer *correct* opinions, usually the authoritative ones. Somewhat paternalistic under Western eyes, such practice has been commonly viewed as a standard of professionalism by Chinese journalists. Note that the Western model of journalism has been increasingly gaining hegemonic power in Chinese academia and the news industry as well. As a result, Western professional journalism packed with liberal notions of reasoning and freedom has been raised and valued. Nevertheless, whether a news report is balanced or not has been not a deep concern for either the media practitioner or the audience during my 20-year journalistic career.

The perceived harmonious relationship between journalism and propaganda in China was not originated from the CPC. It started as early as the emergence of the field of journalism during Republican China. In 1934, the journalism department at Yenching University offered a selective course for juniors and seniors called "Public Opinion and Propaganda" (Wang, X.L., 2010). Two years later, Liang Shichun (1936), the Dean of the journalism department, published the first Chinese textbook on propaganda, entitled "Applied Propaganda Science" (*shiyong xuanchuan xue* 实用宣传学).[2] Under the influence of traditional *xuanchuan* culture, Liang believed that propaganda is an educational tool for the masses to be enlightened. His thought was articulated in the preface of the book: "the purpose of propaganda is to help the masses acquire a clear understanding of an individual, an organization or an ideology, so that they can form a pure, undeceived, and common-interest-oriented public opinion" (Liang, 1936, pp. 1–2). And this by no means is an isolated case.

Meanwhile in Shanghai, Wang Yizhi taught "science of propaganda" at the journalism department at Fudan University. Based on his teaching notes and under sponsorship of the Propaganda Department of the Central Government of the KMT, Wang published his book *Comprehensive propaganda science* in 1944. The credential of the author printed in the cover matter of the book was fascinating, which reads "professor in propaganda science at Fudan University" (Wang, Y., 1944). This is not an issue of translation but a careful deliberation. At the beginning of the book in answering whether propaganda can be considered "science," the author explained the term by referring to the original English word in a rather ambiguous tone. He pointed out that the book's title can be interpreted as "the science of propaganda," "the study of propaganda," or "propagandism (*sic*)" (Wang, Y., 1944, pp. 1–2). What the author was quite certain, however, was that propaganda is a social science because as he reasoned, "if social sciences can be deemed as science, then propaganda is science too" (Wang, Y., 1944, p. 2). As the book's title suggests, this is truly a comprehensive book on propaganda, which was reminiscent of Edward Bernays' book *Propaganda* of 1928. On the other front, the CPC published a 48-page booklet with a similar title, *The science of propaganda*. In this book, Wang Foya (n. d.) pointed out two goals of propaganda: to raise public awareness about politics and to achieve literacy.

From the late 1980s to the 1990s, some Chinese scholars called forth a "new" academic field that they called "*xuanchuan xue* 宣传学," literally, "science of propaganda." Many books with "science of propaganda" in their titles were published in China during this period (see for example, Li, D., 1992; Wang, X.H., 1994; Zhang & Qian, 1992; Zheng, 1987). Nevertheless, these books read more like guidebooks on how to improve propaganda techniques than serious academic work by Western standards. In a book entitled, *The science of mass propaganda*, Gu (1999) put it bluntly that the book aims for training young propagandists. Some Chinese communication scholars insisted that it is just a matter of translation; they viewed "science of propaganda" as "propaganda studies." However, it is evident that at least some propounders held that propaganda can be a science. Cao (1987), for example, considered propaganda the same kind of social science as economics but yet to be developed. Cao's enthusiasm for what he saw as a promising academic field does not come from a vacuum. Cao (1987) argued that "propaganda is a social practice and developed in accordance with the development of the human society and predictably, will continue to make a great impact on our society" (p. 46).

The everlasting debate on the tangled relationship between propaganda and journalism has inevitably brought communication into a triangle. At the center of the triangle is dissemination. The triangle showcases the cultural gap and internal multifarious struggles between the West and China in communication studies. It also manifests itself, as I have shown elsewhere, in what some Chinese scholars call the "Chinese characteristics of communication" (Lin & Nerone, 2015). Let me briefly reiterate it.

The field of communication is translated as *chuan bo xue* 传播学 in Chinese, literally, the "science of dissemination." From "communication" to "dissemination," the boundary drawn between the West and China was fixed. It is of great theoretical complexity, and certainly not a merely linguistic problem. It is, indeed, a cultural gap camouflaged by translatability. Nonetheless, this gap has been rarely, if at all, noted by both Chinese and Western scholarship. The English word "communication" emphasizes *the activity* of exchanging or sharing information, where the central power is usually not implied. That is, everyone can communicate with everyone else. The Chinese word *chuanbo* 传播 or "dissemination," on the other hand, centers on the *ability/capacity* to distribute information to wider audiences, where power is, whether explicitly or implicitly, always assumed in the hands of the one who disseminates. Chinese communication scholar Li Bin (1990) argues that "mass communication is from propaganda, focuses on propaganda, and applies to propaganda, therefore, is the science of propaganda *per se*" (p. 77). For the same reason, another Chinese communication scholar Liu Jianming (2011) claimed that the CPC's propagandists were indeed the original creators of Communication Study in China.

The Cultural Revolution is regarded as a turning point in the history of *xuanchuan* by many. Chinese scholars have noted that it left an unpleasant taste in the public perception of *xuanchuan* (Cao, 1987; Wang, Z., 1982). Ge (1984) argued that journalism was turned into "pure" propaganda during the Cultural

Revolution. Liu (2013) claimed that the notoriety of the Cultural Revolution propaganda combined with bombardments of commercial propaganda after the economic reform resulted in the public dislike of the word *xuanchuan*. Yet, some Chinese writers had acknowledged that the unfavorable attitude toward *xuanchuan* appeared much earlier. Lu Xun 鲁迅 (1881–1936) (2005) pointed out in the 1930s that the word *xuanchuan* had been abused and trashed by Chinese social elites and finally became a nickname for "lie" during the KMT regime. But this is rather a case of abuse of the term, not its significations. Still, in today's China the neutral way of using *xuanchuan* is dominant, from public discourse to government documents. Nothing suggests a connotative change that happened or is happening except the "dislike." To gain the flavor of *xuanchuan*, "seeder" can be a master metaphor. Historically, *xuanchuan* was used for planting "seeds" of good deeds. The Chinese character *xuan* as in *xuan-chuan* first appeared at least in B.C. 990 in a phrase "*xuanqi dexing* 宣其德行," (propagating the virtue) (Wu & Wu, 1695/1995). In his lecture of 1935, Mao Zedong (1991) regarded the Long March as a triumph of *xuanchuan*, naming it the "seeder."

Xuanchuan and propaganda share some commonalities. They are not mutually exclusive categories and the nexus cannot be ignored. First, propaganda is included in *xuanchuan* (Figure 2.1). This is to say, the sinister connotation along with the negative association "package" of propaganda can also be found in *xuanchuan*. For instance, the government's intervention in news reporting can be seen as both propaganda and *xuanchuan*. Secondly, both *xuanchuan* and propaganda intersect with worldview and ideology. The difference is that *xuanchuan* always emphasizes education, whereas propaganda in the West does not. Martin (1929) reckoned that "education aims at independence of judgment. Propaganda offers readymade opinions for the unthinking herd" (p. 145). Randomly asking writers who somehow

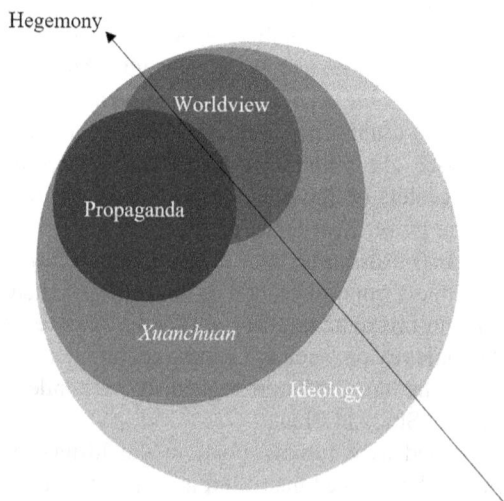

Figure 2.1 The propaganda and *xuanchuan* nexus. Created by the author.

wrote about propaganda, Lapiere and Farnsworth (1936) found that one thing those Western authors held in common was that propaganda is needed to be separated from education. Note that hegemony in Figure 2.1 is interpreted as a place or a direction that all propagandas, *xuanchuan*, worldviews, and ideologies are headed for. That is, all of these "things" are aimed for hegemonic power.

The connection between propaganda and *xuanchuan* also lies in the historical subtlety. Few Western scholars have treated propaganda as *xuanchuan*. For example, Black (2001) argued that like free market to a capitalism system, propaganda is necessary to a fully functioning democratic society. Black (2001) envisaged propaganda as the marketplace of ideas somewhat like the contemporary Western notion of journalism:

> In a politically competitive democracy and a commercially competitive free enterprise system, mass communication functions by allowing a competitive arena in which the advocates of all can do battle. What many call propaganda therefore becomes part of that open marketplace of ideas; it is not only inevitable, but may be desirable that there are openly recognizable and competing propagandas in a democratic society, propagandas that challenge all of us—producers and consumers—to wisely sift and sort through them.
>
> (p. 135)

Likewise, Edward Bernays did not see propaganda as an intrinsically undemocratic instrument. Contrarily, Bernays (1942) enthusiastically stuck up for the idea of using propaganda to strengthen democracy by word-of-mouth methods and public education programs, all of which were strikingly reminiscent of the core of *xuanchuan*. John Perry also valorized the role of propaganda in advancing US democracy and his method reminded us of what Mao achieved during the Chinese Revolution. Specifically, Perry (1942) suggested that government propaganda and channels in the United States can be effectively used to spread information, facts, and ideas, to provoke discussions, and to facilitate education by outreaching farmers, labors, housewives, and the youth.

Why in China?

One crucial question that remains to be asked is why the propaganda culture in China did not also emerge in the West. While this question cannot be answered sufficiently without full inquiries into Chinese history, culture, politics, and philosophies, it is too important to be ignored. For the answer, I look into two crucial factors: religion and culture. This is, by all means, a daunting task. What I offer here are more assumptions than assertions. At the beginning of the *Critique of Hegel's philosophy of right*, Marx (1970) makes a point that any criticism should start with the criticism of religion. Ricoeur (1986) remarked that the critique of religion provides us a model for a critique of ideology. But China does not have *a* religion; it has many religions, yet none considered a state religion. I think that this matters greatly to Chinese communication and so does the *xuanchuan* culture.

My first assumption is that the *xuanchuan* culture in part operates as a substitute for the missing state religion to unify the nation in a quasi-religious way. One thing that religion and propaganda share in common is that both encourage and demand devotion. *Xuanchuan*, derived from Confucian teachings, can be seen as a kind of "rationalized religion" like Confucianism itself, "more abstract, more logically coherent, and more generally phrased" compared to traditional religion (Geertz, 1973, p. 172). Perhaps, the most auspicious place to look for the role of religion for Chinese communication is in Weber's book *The religion of China*. Weber (1964) repeatedly used the word "decisive" throughout. For Weber, the absence of unified religion in China was sociologically decisive. This leads Weber (1964) to assume a kind of ideological polarization unique to traditional China: lacking "a particular mentality" on the one hand and abundant in Chinese ethos on the other (p. 104). The mentality that Weber referred to is equivalent to the Protestant ethic and the spirit of capitalism (Weber, 1958). According to Weber, the polarization had two profound ramifications that have devised the social structure of China. First of all, capitalism failed to develop in China and communities in the Western religious sense did not exist. The second consequence, perhaps more enlightening, was that ideological wars became very frequent. Weber (1958) argued that the clash between the Confucian literati and the anti-literati Taoists was "decisive" for the "structure of Chinese politics and culture" (p. 46). Such clashes left deep traces on the *xuanchuan* culture. Another interesting point Weber (1964) made, among others, is that unlike in the Middle Eastern states, there was no influential prophets in China, suggesting that the Confucian "ritualist," "literary officeholder," and "the emperor" acted as proxy for the prophet. Like missionaries working in an archetypical propaganda system, the three players did a similar job that Weber (1964) would describe as the taming of the masses. This is *xuanchuan per se*.

Xuanchuan is the way of communication and life that keeps Chinese people searching for guidance for life, *raison d'être*, and other meanings, which has earned *xuanchuan* a pseudo-religion-like connotation. Today in China, it is still the case that everyone propagandizes everyone else in their daily life or on the Internet about many ideologies and worldviews, including bourgeois lifestyle, Confucianism, Americanism/anti-Americanism, patriotism, Maoism/anti-Maoism, many forms of Buddhism, etc. Without the presence of a unified religion, people treat ideology as religion when disseminating it as if it were an absolute truth. As in a religious world, groups propagandize all sorts of ideologies with the same conviction that what they do is for the common good.

My second assumption valorizes the role of collectivism as a cultural pattern in forming the *xuanchuan* practice. This is in line with the theory of dimensions of cultures proposed by Geert Hofstede. In a large survey conducted in the late 1960s and early 1970s about the values of people working at IBM from 50 countries, Hofstede (1991) identified "power distance," "collectivism versus individualism," "femininity versus masculinity," and "uncertainty avoidance" as the four basic areas of common problems worldwide. He called those "four dimensions of cultures," together a four-dimensional (4-D) model. Hofstede (1991) postulated that the 4-D model can help map out a country's culture. Later, based on Confucian dynamism, Hofstede advanced his model by adding a fifth dimension labeled "long-term orientation." The Hofstede model has been extended and applied to many areas of the social sciences and generated a growing body of literature, becoming a "doctrine" for intercultural communication (Minkov & Hofstede, 2011). Applying the Hofstede model, empirical studies, mainly conducted in the field of business management, have demonstrated a cultural gap between China and the United States: China ranks high in "power distance," "collectivism," and "long-term orientation," whereas the United States ranks lower in these areas (Shi & Wang, 2011). As a result, what some may call the "Chinese mind" emerged. Characterized as exhibiting a high degree of authoritarianism, collectivism, and long-term orientation, the Chinese mind speaks a lot to the aforementioned four characteristics of *xuanchuan* practice. *Xuanchuan* aims to integrate individuals and social groups into a Confucian harmonious society where the authoritative power and rules are to be respected and obeyed. It is also based on collectivistic culture, mass mobility, and social networking, the latter two characteristics work powerfully for producing propaganda. I do not mean to suggest, simplistically, that human mobility and social communication are not effective in an individualistic culture. Rather, I want to address that, compared to the China's case of "everyone talking to everyone else," propaganda in Western individualistic countries has a higher degree of mass media dependency. Moreover, *xuanchuan* generally seeks to achieve long-term effects through inculcation and distillation rather than short-term shocking impact (e.g., brainwashing). This is why in practice education has been intricately intertwined with propaganda. "A sensational finding," in Hofstede's description, of this scholarship was the correlation between Confucian values and economic growth (Hofstede, 1991; Hofstede & Bond, 1988). It demonstrates that culture in a certain form of national values can boost its economy. I have argued elsewhere that China's propaganda has centered on the economy and contributed to the nation's economic growth even during the turbulent period of the Cultural Revolution (Lin, 2013). Linking Confucian dynamism to *xuanchuan* practice, at a minimum, would shed some light on the puzzle of why China's economy has continued to grow without liberalizing its propaganda system. Nevertheless, it should be noted that there is a potential to overapply Hofstede's model to solve the complicated Chinese communication problem. Again, I herein only mean to invoke the normative power of Hofstede's model in order to stimulate thought on the mass behavior of *xuanchuan*. Any deductions more than that would be exaggerated and invalid here.

The problem of representation

What should be clear by now is that the conceptual morass of propaganda is largely due to its intricate connection to ideology. To a large extent, the problem of propaganda is the problem of ideology. Compared to propaganda, the term "ideology" is less pejorative but overwhelmingly inclusive as it deals with almost everything about human minds. This leads Ricoeur (1986, p. 138) to argue that the field of ideology is "so wide," the field of science "so narrow." The problem also goes to the famous "Mannheim's paradox": if everything coming from our minds is ideological, how is it possible that a theory of ideology can be itself non-ideological? Now the subject becomes a philosophical minefield. To cross the realm of ideology without being trapped into obscure jargons, I take the idea of "Representation" as the central axis of my discussion. I capitalize the term "Representation" to designate its philosophical connotations, and hence to differentiate it from the common use in communication study particularly as media representation. Discussing ideology with respect to Representation, I want to drag the topic into the field of communication, linking it to propaganda. Nonetheless, this is not an arbitrary move. At the heart of conceptions of ideology by many thinkers is the Representation *per se*, whether called "a system of representations" by Althusser (1979), "distortion" by Marx (1998), "symbolic interaction" by Habermas (1971), "sophistic or rhetorical structure" by Weber (Ricoeur, 1986), or "symbolic mediation of action" by Ricoeur (1986).

It has been said that a sophisticated discussion of ideology should start with Marx and his *The German ideology* (Eagleton, 1991; Ricoeur, 1986). For Marx, ideology is a distorted Representation of praxis. It is not opposed to science and truth, so it is not an untruth or a lie. To Marx, ideology is analogous to the inversion of an image (*camera obscura*) in a sense that it is symbolically mediated through communication. Ricoeur (1986) also noted that ideology is "representation and not real praxis" (p. 77). Habermas (1971) likewise points out that praxis has two dimensions, namely, instrumental action and symbolic interaction. The later dimension, as he suggests, is where "the configurations of consciousness" takes place, referring to Marx's notion of ideology as "manifestations" of this process (Habermas, 1971, p. 42). Here, "manifestations" is merely another way to say "representations." Habermas goes on, arguing that labor in its commodity form is ideology because it represents the social relation of the production.

Another way, perhaps an easier one, to comprehend Marx's abstract idea of ideology is to look at the real person who runs ideology: the ideologist. In a marginal note Marx (1998) remarks that the first form of ideologist is the priest. This reminds us of the missionary in the prototypical propaganda network. In Marx's account, ideologists were given birth through the division of labor, particularly the division of mental and material labor in the ruling class. A thinker of the ruling class, the ideologist "make[s] the perfecting of the illusion of the class about itself their chief source of livelihood" (Marx, 1998, p. 68). This is to say, ideologists worked as some kind of media practitioner to serve the ruling class in their capacity of mediating representations. This image of the ideologist strikingly parallels the public imagination of the propagandist insofar as the primary role for both the ideologist and the propagandist is to make up desirable images for the ruling class.

Weber's idea of ideology as "legitimation" is one step closer to propaganda. For Weber (1946; 2004), Marx's notion of distortion is legitimation *per se*. In other words, ideology is the legitimated Representation. According to Weber, the way to legitimate Representation is through coercion in forms of "sophistic or rhetorical structure" (Ricoeur, 1986, p. 195). Therefore, Representation turns out to be the legitimation of political authority by coercive and deceptive means. Now a demonic image of propaganda looms. Eagleton (1991) suggests that the process of legitimation involves at least six strategies:

> [P]romoting beliefs and values congenial to it; naturalizing and universalizing such beliefs so as to render them self-evident and apparently inevitable; denigrating ideas which might challenge it; excluding rival forms of thought, perhaps by some unspoken but systematic logic; and obscuring social reality in ways convenient to itself.
>
> (Italic in the original, pp. 5–6)

Through this legitimating process, a certain set of values becomes something that Bourdieu (1977) would call doxa, except it is manufactured. It reminds us of "manufactured consent" in Edward Herman and Noam Chomsky's (1988/2008) description. Ideology and propaganda are so intertwined in the realm of legitimation that Eagleton (1991) takes Mao's propaganda slogan "the West is a paper tiger" as an example of ideology (p. 26).

Contrasting to Weber, Ricoeur developed a neutral thesis of Representation, viewing ideology as integration/identification for social groups. Ricoeur (1986) claimed that ideology helps social groups to be integrated in both time and space. He argued that "[i]deology preserves identity, but it also wants to conserve what exists and is therefore already a resistance" (Ricoeur, 1986, p. 266). To illustrate how ideology integrates a social group in a non-pejorative way, Ricoeur took individualism as an example of the ideology of the United States. In that regard, Ricoeur was rather speaking of the US culture than an abstract ideology. This is because treating ideology as an integrative force by means of symbolic interaction is, in another way, showing an example of how culture works. A bearer of integrating power, Representation is now taking us to the changing terrain of culture, where everything communicates everything else. To be specific, we need to take Marx's notion of distortion and Weber's legitimation of authority into consideration to comprehend Ricoeur's idea of integration. It should be noted that the three ideas from the three authors are rather three dimensions of the same conceptual construct, the Representation, not three distinctive conceptions. I think Ricoeur (1986) would agree since he, interestingly, refers to "distortion," "legitimation," and "integration" as the three "functions" or "roles" of ideology in a loose manner (pp. 265–266).

I found Ricoeur's notion of ideology as integration particularly helpful for the study of Red Tourism at least in two respects. First, it prioritizes a neutral reading of ideology, which has given me a disposition to consider what Red Tourism delivers not something deceptively manipulative. Coincidently, Ricoeur's ideology integration echoes Ellul's (1965) integration propaganda in the way that both are thought to unify social groups. Secondly, Ricoeur (1986) puts ideology and utopia in the same framework of Representation, pointing out that the two seemingly opposing phenomena work together constructing what he refers to as "cultural imagination" (p. 1). What Ricoeur suggests here is that cultural imagination is not merely a cultural product, but also an ideological one.

Hegemony is what Representation is intended to achieve. Theorized by Antonio Gramsci, hegemony was originally treated as a ruling power just as force (Bates, 1975). Gramsci (1988) wedded ideology and propaganda into a phrase "ideological propaganda" to address the hegemonic power of a certain type of propaganda such as Fascist propaganda and the Communist Party's propaganda. Hegemony assumed a nefarious image partly because of its association with dictatorship. But this is rather a misinterpretation of Gramsci's idea than his own. Gramsci used hegemony somewhat in a neutral way. In discussion of Italian popular culture, for example, Gramsci (1988) referred to the Italian readers' preference of foreign writers as undergoing the "moral and intellectual hegemony of foreign intellectuals" (p. 367). What Gramsci is saying here is that the foreign writers wrote something closer to the daily life of the Italian people than what Italian indigenous writers did. Hegemony goes in many forms (cultural, moral, intellectual) other than ideology. For this consideration, Eagleton (1991) has argued that hegemony "*includes* ideology, but [is] not reducible to it" (emphasis in the original, p. 112).

Raymond Williams also connected hegemony to propaganda in an inexplicit way, though he had very little to say directly about propaganda. Williams (1980/1997) considered hegemony a dominant system of practices, meanings, and values, which appears in any period. Williams illuminated the way hegemony is transformed into the dominant culture through a mechanism that he referred to as "the selective tradition." Characterized by intended and systematic inclusion, emphasis, re-interpretation, omission, and exclusion, the selective tradition is the Representation that mediates and is mediated by the past and the present. Williams (1980/1997) goes on to note that it is not easy for us to get rid of the dominant culture, because it is not something the ruling class imposes on us but something "built into our living" (p. 39). This can be seen as a different expression of the idea of propaganda as a social space, produced and productive.

Neighboring ideology and hegemony is another Representation called "worldview." Worldview is the Representation of the world in people's minds. Geertz (1973) described it as the "picture of the way things in sheer actuality are, their concept of nature, of self, of society" representing people's "most comprehensive ideas of order" (p. 127). For Williams (1980/1997), worldview is "the organized way of seeing the world" (p. 24). By virtue of this, scholars tend to use ideology and worldview interchangeably. For instance, a dominant worldview is a dominant ideology. Habermas (1971) extends worldview to connote a certain system of knowledge. He refers to science as the natural/scientific worldview, and others include "philosophical worldviews" and "religious worldviews." Nevertheless, as Representation with a capital "R," worldview always comes into being with its hegemonic power.

The *xuanchuan* model

Drawing from the cultural practice of *xuanchuan*, I propose a neutral and broader conceptual construct of propaganda for the current study. Herein I name it the *xuanchuan* model, characterized by the goal of social integration, educational instrumentality, priority of human mobility, and social networking method. It assumes that propaganda is a social space through which ideology is to be disseminated, propagated, promoted, and reproduced in the form of Representation for the purpose of generating hegemony. It also assumes that propaganda and ideology coexist in a symbiotic space, meaning that they are mutually dependent phenomena. On the one hand, ideology as Representation does not run things by itself, and on the other hand, propaganda without ideological signification is unthinkable.

But we have an ingrained habit to think propaganda as certain ideologies and to separate propaganda and ideology from their social space. The folklore has been that anything coming from the Communist Party, the propaganda department, the *People's Daily*, or the Chinese political leaders is propaganda, regardless of textual property and social context. In other words, political ideology itself is deemed as propaganda. But how do other ideologies, say, liberalism, neoliberalism, racism, colonialism/post-colonialism, cultural imperialism, individualism, orientalism, utopianism, operate? Racism nowadays does not have its formal organizations, institutions, media, leaders, and nor do the other aforementioned ideologies. No professionals work to generate "publicity" for those. Propagation of these ideologies largely hinges on popular culture, economic activities, and sometimes, military actions (think about colonialism). This implies two corollaries.

First, popular culture is a popular site of propaganda. As Representation, ideology is embedded and embodied in films, TV dramas, songs, fashions, news stories, commercials, live concerts, shows, national/international sports events, and so forth. Propaganda of any sort has a marked tendency to mobilize popular culture for delivering intended ideological messages. Jowett and O'Donnell (2012) have noted that "mass propaganda is now largely practiced through trade, travel, and exchange of culture and scientific and sporting achievements and not through warfare" (p. 288). It is also evident in the invention of revolutionary opera and mass production of revolutionary songs during the Cultural Revolution, and also in many Disney shows. Here, again, I opt for the neutral connotation of propaganda based on a neutral conception of ideology as Representation. In a sense, fascism is an ideology, so are Western liberalism and communism. Likewise, spreading out communism is propaganda and promoting individualism is propaganda too. Neither am I suggesting that propagandistic shows of Disney are produced to manipulate the audience's opinions, whatever they are, nor arguing that the people who watch those shows would necessarily be victimized by such propaganda. Rather, my point is that popular culture in any form works for propaganda just like other propaganda machineries/ organs. By the same token, Jowett and O'Donnell (2012) consider Thomas Paine (1737–1809) the "first great propagandist of the American Revolution" along with other American propagandists including Samuel Adams, Benjamin Franklin, and Thomas Jefferson (p. 80). As a form of popular culture, tourism is no exception. This leads me to argue in the next chapter that tourism can be seen as a propaganda system.

The second corollary is that propaganda includes both discursive communications and non-discursive networks. This is to say, propaganda is a social space. It is produced by and producing diverse discourse such as histories, collective memories, nostalgia, sentiments, national identities, and non-discourse such as economy and praxis. This can, at least in part, explain why there are no specific media and institutions responsible for racism, but racist groups have constantly grown fast worldwide. I will articulate and elaborate the idea of propaganda as a social space and how it speaks to Red Tourism in subsequent chapters.

In what remains of this chapter, I will try to think through the Chinese characteristics of propaganda by moving back and forth and extending what I have touched on earlier. I would argue that China is a propaganda state. A seemingly familiar way of framing the CPC's media politics by Western mainstream news media; however, my definition of "propaganda state" is different. My intent and exploration on the subject are based on a school of thought on propaganda, the *xuanchuan* model, and most of all, a culture of propaganda. As noted earlier, all of these had existed long before the CPC came into power. China is a propaganda state on the ground that propaganda is a Chinese characteristic form of governance and a Chinese characteristic way of mass communication.

The Chinese characteristic form of governance

The imagined communist society would have been *déjà vu* for some Chinese people long before the advent of Marx's *The Communist manifesto*. Confucius depicted that community as the "world of great harmony" (*shijie datong* 世界大同) where everyone loves each other, every family lives and works in peace, no disputes, and no wars. The sharp divide between Confucian cosmopolitanism and Marxist communism lies in the pathways to achieve them. While Marx saw the proletariat revolution, Confucians advocated the sort of massive, persuasive, ideological education, believing that it would lead to the formation of the cosmopolitan society. That pathway is *xuanchaun*/propaganda.

Propaganda was also believed to be the optimal way of governance. The propaganda state was envisaged as the ideal one that is governed by effective and benevolent communication between the ruling class and the ruled, not by coercion. This view was not limited to Confucians but shared by many schools of thought. For example, Guan Zhong 管仲 (723–645 B.C.E.), a key figure of Legalism (*fajia* 法家), made it explicit in *The Art of the Mind* (心术) that "to govern a state was to govern the people's minds," (心治是国治也) and it was "not done by punishments and coercive sanctions" (操者非刑也). The combination of a virtuous ruler, an effective mass communication system, and the means of self-improvement was of crucial importance in running such a propaganda state. The rationale goes like this: the ruler would provide the right kind of guidance to mobilize the ruled through mass communication networks for their self-improvement. Traditional cultural forms in China, in line with this propaganda reasoning, were aimed at disseminating worldviews and lifestyles. Symbolic meanings of traditional Chinese paintings and poems, no matter how sophisticated, normally can be decoded by associating them with a certain kind of worldview or ideology. But the Confucian idea of governing minds through propaganda did not go unchallenged.

Three contentions occurred. One challenge came from the monarchs in imperial China, one from the liberals in Republican China, and one from the Communists in Socialist China. The first crisis arose when Emperor Qin Shi Huang 秦始皇 (259–210 B.C.E.) ordered the burning of Confucian classic books and the historical books of other states, known as "burning books and burying Confucian scholars alive" (焚书坑儒). Emperor Qin saw Confucianism as a heterodox spirituality. The second anti-Confucianism campaign appeared during the May Fourth Movement of 1919. Given the nickname of "*kong jia dian* 孔家店," or "Confucius and Sons" by Hu Shi 胡适 (1891–1962), a student of John Dewey at Columbia, the Confucian school of thought was believed anti-democratic by the Chinese liberals back then. The last attack on Confucianism took place in the "Criticizing Lin and Confucius Campaign" during the Cultural Revolution when the CPC deemed it counterrevolutionary. Retrospectively, it is clear that what the emperor, the liberals, and the Party members fought against was not the practice of propaganda, but rather a matter of replacing one propaganda with another, whether despotism, Western liberalism, or Maoism. In that sense, propaganda as a form of governance was generally desirable and rarely challenged throughout Chinese history.

This Chinese characteristic form of governance has been gradually recognized in the field of contemporary China studies. Nevertheless, influenced by the hegemonic Western model of propaganda, scholars have unanimously tended to think of such propaganda form of governance as a new propaganda maneuver of the party. Associating it with "cultural governance," "cultural-*cum*-nationalist propaganda," or "cultural nationalism," for example, Perry (2013) described the propaganda practice as the strategic deployment of symbolic resources by the party to deal with its legitimacy crisis in post-1989 years. In a weblog post, Cheek (2016) has pointed out that the "Chinese model of propaganda" is the way China works. Cheek pushed the date back, but not any further, seeing the Chinese model as an invention and later, a tradition of the party. While this may explain, to some degree, why propaganda matters so much to the CPC, it cannot explain why propaganda also matters so much to the Chinese public. For example, why have Chinese netizens taken television propaganda dramas featuring the Chinese Civil War and the War of Resistance Against Japanese Aggression so seriously that they kept demanding stricter censorship from the government to filter out dramatic martial arts sequences and lengthy sexual scenes? For the sake of entertainment, these scenes are exactly what television dramas are all about. Likewise, why did millions of Chinese tourists choose Red Tourism at the time when their capacity for traveling freely and internationally was so high? And why has their Red Tourism experience been consistently and markedly positive? These questions cannot be explained by a theory of brainwashing. Nor is the explanation of a political party's strategy sufficient. It becomes even more confusing and problematic as the debate on the distinction between propaganda and journalism in Chinese academia and among media practitioners continues.

The Chinese characteristic way of mass communication

Some Chinese people see journalism quite differently. For them, the role of the press is not merely an information provider, but more importantly, as intellectual guidance. If professional journalism in the United States can be characterized by a balanced model, the Chinese model of journalism can be, in my view, identified with a "biased" mode. But bias, like many other terms in communication study, is highly misleading. For one thing, bias only exists by comparison with other biases. In that sense, balance in professional journalism is, too, a bias. For another, as communication scholar Dan Schiller has spelled out in his lecture, there are bad biases and there are also some good biases. For example, the bias of some races being inferior to others versus the bias that the world would eventually become perfect and everyone is happy. Perhaps, for the majority of Chinese journalists, the ultimate goal of professional journalism is to *propagate* good "biases," not to balance good ones with bad ones. But judging what is good or bad is both moral and ideological. It is the sort of agency that a Chinese journalist is believed to produce good journalism. Therefore, journalists themselves play a decisive role in this "biased" model of journalism, not their professionalism. The audience expects and demands that journalists show their good judgment with which they can choose to agree or disagree. Balanced journalism without a marked "bias" is usually interpreted as insignificant to political discourse. Chinese newspeople rarely tend to hide the propagandistic nature of their work. The *Metro Express*, a metropolitan daily newspaper of the news group I used to work in, sports the rather audacious slogan of "To Influence Those Who Have the Greater Influence."

The "biased model" grew out of a cultural tradition in China. Neither was it a result of the party's media politics, nor did it stem from a particular journalistic practice. It originated from a shared tradition with other Chinese cultural forms, as I have argued here and there, that are largely aimed at mass ideological education for social integration. This tradition remains so deeply ingrained in today's Chinese culture industry that it became a robust criterion for evaluating a cultural product or work. The ideological function of cultural products has been enshrined to an unbearable degree. For example, *xiaopin* 小品, China's equivalent of TV standup comedy, are oftentimes the opposite of humor, taking on emotional and grave matters. The rationale is that a serious, melodramatic monologue is more ideologically evocative than a purely entertaining and humorous one. In other words, bringing the audience to tears would teach them much more than simply bringing out laughter. Tourism has parallels to this. Like their ancestors, many contemporary Chinese tourists still hold the belief of what I refer to as "serious tourism": that good tourism should be treated as a serious matter and planned out with the goal of personal growth and self-improvement. Red Tourism, for better or worse, fits that description perfectly; the ends of Red Tourism are of understanding and contemplation of revolutionary struggle, despite being ostensibly recreational travel.

Another, perhaps more intellectual paraphrasing of the "biased" model is the *xuanchuan*/propaganda model. Historically speaking, Chinese journalism originated from propaganda and was aimed for propaganda. Starting in the early 1900s, Chinese newspapers became a mouthpiece for various political activist groups, namely, the reformists, the revolutionaries, and the constitutionalists in the course of China's political modernization (Fang, 2000). *Shibao* 时报is a good example. Instead of subverting the Qing government and overthrowing the imperial regime by violence, Liang Qichao found the propaganda power of the press, creating *Shibao*, a daily newspaper in 1904 in Shanghai. Not only did *Shibao* propagate a constitutionalist reform agenda, but also it promoted a sort of modern lifestyle. Nevertheless, the reformist newspapermen did not stand against Confucianism as some supposed. Rather, as Judge (1996) has pointed out, they "infused the Confucian tradition with new elements and transformed foreign ideas to conform to familiar cultural constructs" (p. 4). In other words, what they did was to mobilize Confucianism for their own political agenda. Does this practice remind of the rise of the Confucius Institute worldwide?

Sadly, the long-lasting debate on the relationship between journalism and *xuanchuan*/propaganda in China and enduring propaganda research on contemporary China in the West excluded a very important chapter of a culture of propaganda. In China, the hope of creating what some early Chinese scholars called "propagandism" or "science of propaganda" was dashed by the hegemonic Western model of journalism alongside the hegemonic Western model of propaganda. In a major Chinese scholarly work on propaganda, Liu Hailong (2013) named the American model of propaganda – marked by public relations, crisis management, and spokesmanship – the "scientific model" (p. 5), which he thought that China should adopt. Coincidentally, balanced journalism has gained some currency in Chinese journalism students and some elite news media, even though the legitimacy of the golden rule of balance in Western professional journalism has been continuously questioned by journalism scholars in the West, whether referring to it as "addiction" (Nerone, 2015) or "strategic ritual" (Tuchman, 1972). Nevertheless, the spirit of Chinese journalism, which rose from a much longer Chinese literary tradition emphasizing the propagation of morals and virtues, still holds Chinese journalists together. Mainland Chinese journalists calling themselves "the ones doing propaganda work" can only be explained in this spirit of journalism.

It is this deep-seated belief in the benevolent power of *xuanchuan*/propaganda that persisted down through the centuries in China in spite of, almost in defiance of, the prevalence of the malevolent power of propaganda elsewhere in the world. It is also because of this belief that each day millions of Chinese tourists follow the Red routes in the hope of gaining insights for making better sense of the past, the future, and perhaps more importantly, the link between the two.

In this digital age marked by the universal ideology of trendy and ubiquitous swarming followers on the Internet and social media platforms, the brand of "propaganda state" is no longer exclusive to China or a handful of other sovereign states. Never was propaganda so pervasive in everyday life in human history as today. Like how the Chinese people see propaganda, it always cuts two ways: more possible democratic communication yet more controllable.

Conclusion

The problem of propaganda is that it has too many problems.

The insurmountable problem of propaganda is its ideological orientation. It assumes that propaganda is a feature of certain ideologies, political/social systems, whereas others are very much immune to it. Although many media scholars and critics have said that propaganda exists in all human societies, it has not swayed the social imaginary of propaganda as the special product of certain sets of historical periods, nations, and political leaders. This is what people have been constantly fed by popular culture where propaganda – often coming disguised in the form of foreign evildoers, government moles, or highly framed international news coverage – is ideologically determined. It has fostered a massive illusion that propaganda is the issue of "Others." It follows that propaganda in developing countries exerts greater influence than in Western democratic countries, thusly much more noxious.

This, in turn, defines the evaluative nature of the term, which is a perennial problem of propaganda. In other words, the word "propaganda" evokes conation rather than cognition. Case in point, look at the "distinction" between propaganda and counterpropaganda. It has been a tradition in the United States to call spreading disfavored ideologies "propaganda", such as Nazi propaganda in WWII, Soviet/Communist propaganda in the Cold War, and Jihadist/ISIS propaganda in the war against terrorism, whereas US propaganda is renamed as "counterpropaganda." Herbert Romerstein (2009), a former director at the propaganda institution of the USIA, refers to counterpropaganda as "carefully prepared answers to false propaganda" (p. 137). It is clear that in his definition, US manipulation is rephrased as "carefully prepared answers" to connote something rather positive. This is also true for many typologies of propaganda. For example, Ellul (1965) divided propagandas into many binaries based on the propaganda producer's intention and the receiver's inclination. Even for a neutral thesis of propaganda, the author is always required to determine whether the propaganda is good or bad, say, is Red Tourism bad propaganda or good propaganda? The evaluative nature has rendered the study of propaganda more political and less academic.

The overemphasis of vertical influence is another problem. From the hypodermic needle model to the two-step flow of communication, theories related to the study of propaganda have directed much attention to hierarchal power structures and the powerful people. To a great extent, propaganda has been viewed as the event of the state, say, the most powerful minority bamboozling the helpless majority. Relatively, how propaganda is (re)produced and disseminated in the private has received scant attention. Put very simply, the horizontal influence of propaganda has been largely overlooked. There are still a lot of other problems including textuality dependency, short-term effect primacy, etc. Ultimately, the problem of propaganda has become a deterrent to prevent scholars from discovering the complexity and multilayeredness of propaganda particularly in the field of popular culture studies.

In contrast to the contemporary Western notion of propaganda, *xuanchuan* is a cultural practice in China. This is a little-known fact but has immense significance for studying propaganda against the backdrop of contemporary China. Based on this practice, I propose the *xuanchuan* model. It regards propaganda as a social space, produced and productive. It is considered a neutral framework of propaganda, not only because it considers the ideology that propaganda purposefully tries to propagate neutral, but also the social space of propaganda itself neutral. It is also based on this model that I proposed a propaganda state.

It is worth noting at this point that the *xuanchuan* model is not opposite but supplementary to the Western propaganda model. The Western model makes better sense of propaganda in the state sector, whereas the *xuanchuan* model is aimed at illustrating how propaganda operates in relation to popular culture in the private sector. The distinction between propaganda in the state and in the private is only for the purpose of clarification of the two models. In reality, propaganda as a social space cannot be separated from both the sectors. It should be also noted that *xuanchuan* as a practice of disseminating ideology appears in all modern societies, both in the West and China. The difference is that *xuanchuan* is a culture in China. The seemingly troublesome statement can be illustrated by a comparison of eating rice: it is a cultural practice in China, but Americans also eat rice. Constructed at the intersection of propaganda, space, mobility, and popular culture, the *xuanchuan* model lays the theoretical groundwork for developing another important conceptual construct of this book, where tourism is defined as a popular site and a vehicle of propaganda.

Notes

1 The interplay between transportation and communication in the Western cultural context has been thoroughly studied (e.g. Carey, 1989; Mattelart, 1996), but their work primarily focuses on how transportation facilitates the distribution of mass media. For example, the US postal system helped spread the news (John, 1995). In contrast to the Western scholars, Chinese scholars paid close attention to human mobility beneath the transportation network. Qiu (1993) points out that because of extensive transportation network in the Spring and Autumn Period, wars, leisure and official visiting other countries (*pinxiang*聘享), business travels and political lobbying (*youshi*游士) became increasingly frequent, all of which helped achieve greater human mobility and, ultimately, the ubiquity of *xuanchuan*.
2 In this book, Liang (1936) systematically analyzed a wide variety of social phenomena, media, practices, and techniques associated with propaganda including newspapers, advertising, the Catholic church, college newspapers, education news, public speech, leaflet, etc. Liang suggested that propaganda is not a modern development. However, propaganda techniques became increasingly important to China as the country was undergoing modernization.

References

Althusser, L. (1979). *For Marx*. Vintage Books.
Bates, T. R. (1975). Gramsci and the theory of hegemony. *Journal of the History of Ideas*, *36*(2), 351–366.

Bernays, E. L. (1928). *Propaganda*. Liveright.

Bernays, E. L. (1942). The marketing of national policies: A study of war propaganda. *The Journal of Marketing, 6*(3), 236–244.

Black, J. (2001). Semantics and ethics of propaganda. *Journal of Mass Media Ethics, 16*(2–3), 121–137.

Block, R. (1948). Propaganda and the free society. *Public Opinion Quarterly, 12*(4), 677–686.

Bourdieu, P. (1977). *Outline of a theory of practice*. Cambridge University Press.

Bytwerk, R. L. (1998). The propagandas of Nazi Germany and the German Democratic Republic. *Communication Studies, 49*(2), 158–171.

Cao, F. (1987). Shi lun xuanchuan xue [A note on propaganda studies]. *Sichuan shifan daxue xuebao, 8*, 44–47.

Carey, J. W. (1989). *Communication as culture: Essays on media and society*. Unwin Hyman.

Cheek, T. (2016, August 18). *Xuanchuan* in China: Propaganda as the art of governance. https://blogs.kent.ac.uk/munitions-of-the-mind/2016/08/18/xuanchuan-in-china -propaganda-as-the-art-of-governance/

Chu, G. (1977). *Radical change through communication in Mao's China*. University of Hawai'i Press.

Creel, G. (1941). Propaganda and morale. *American Journal of Sociology, 47*(3), 340–351.

Creel, G. (1972). *How we advertised America*. Arno.

Dai, Y. (1992). *Xiandai xuanchuan xue gailun* [Introduction to modern propaganda science]. Lanzhou daxue chubanshe.

Deng, Z. (1988). Zhongguo gudai xuanchuan huodong chutan [A note on Chinese ancient propaganda practice]. *Shangrao shichuan xuebao, 6*, 41–46.

Dewey, J. (1918, December 21). The new paternalism. *The New Republic*, 216–217.

Doob, L. W. (1935). *Propaganda: Its psychology and technique*. Henri Holt.

Doob, L. W. (1949). *Public opinion and propaganda*. Cresset.

Dowd, D. L. (1951). Art as national propaganda in the French Revolution. *Public Opinion Quarterly, 15*(3), 532–546.

Dymkov, A. (1967). Soviet view of Chinese press propaganda (originally from *Sovetskaya Pechat*, 1966, 12). *The Soviet Press, 5*(3), 1–9.

Eagleton, T. (1991). *Ideology: An introduction. Vol. 9*. Verso.

Ellul, J. (1965). *Propaganda: The formation of men's attitudes*. Alfred A. Knopf.

Fang, H. (2000). Zhongguo xinwen chuanbo shiye bainian [One hundred years of Chinese journalism industry]. *Xinwen sanwei, 12*, 42–46.

Fellows, E. W. (1957). Propaganda and communication: A study in definitions. *Journalism & Mass Communication Quarterly, 34*(4), 431–442.

Fellows, E. W. (1959). "Propaganda": History of a word. *American Speech, 34*(3), 182–189.

Garber, W. (1942). Propaganda analysis-To what ends? *American Journal of Sociology, 48*(2), 240–245.

Ge, C. (1984). Lun xinwen he xuanchuan de guanxi [A discussion on the relationship between journalism and propaganda]. *Xinwen daxue, 2*, 6–11.

Geertz, C. (1973). *The interpretation of culture*. Basic Books.

Gordon, G. N. (1971). *Persuasion: The theory and practice of manipulative communication*. Hastings House.

Gramsci, A. (1988). *An Antonio Gramsci reader: Selected writings, 1916–1935*. Schocken Books.

Gu, C. (2014). Zhong xiao: mingqing shehui de hexin jiazhiguan [Loyalty and filial piety: The core value of the Ming and Qing societies]. *Difang wenhua yanjiu, 2*, 23–29.

Gu, Z. (1999). *Dazhong xuanchuan xue* [The science of mass propaganda]. Guangdong gaodeng jiaoyu chubanshe.

Guo, Z. (1985). *Xianqin zhuzi xuanchuan lungao* [Manuscripts on propaganda by pre-Qin philosophers]. Fujian renmin chubanshe.

Habermas, J. (1971). *Knowledge and human interests*. Beacon Press.

Herman, E. S., & Chomsky, N. (1988/2008). *Manufacturing consent: The political economy of the mass media*. Pantheon.

Hofstede, G. (1991). *Cultures and organizations: Software of the mind*. McGraw-Hill.

Hofstede, G., & Bond, M. H. (1988). The Confucius connection: From cultural roots to economic growth. *Organizational Dynamics, 16*(4), 5–21.

Horkheimer, M., & Adorno, T. (2002). *Dialectic of enlightenment*. Stanford University Press.

Hosterman, C. A. (1981). Teaching propaganda. *Communication Education, 30*(2), 156–162.

Hwang, K. K. (1999). Filial piety and loyalty: Two types of social identification in Confucianism. *Asian Journal of Social Psychology, 2*, 163–183.

Jowett, G. S. (1987). Propaganda and communication: The re-emergence of a research tradition. *Journal of Communication, 37*(1), 97–114.

Jowett, G. S., & O'Donnell, V. (2012). *Propaganda and persuasion*. Sage.

Judge, J. (1996). *Print and politics: Shibao and the culture of reform in late Qing China*. Stanford University Press.

Katz, E., & Lazarsfeld, P. F. (2006). *Personal influence*. Transaction.

Kenez, P. (1985). *The birth of the propaganda state: Soviet methods of mass mobilization, 1917–1929*. Cambridge University Press.

Lapiere, R. T., & Farnsworth, P. R. (1936). *Social psychology*. McGraw-Hill.

Lasswell, H. D. (1927). The theory of political propaganda. *American Political Science Review, 21*(3), 627–631.

Lasswell, H. D. (1948). Propaganda. In *Encyclopedia of the social sciences, 11*. MacMillan.

Li, B. (1990). Chuanboxue ji xuanchuan xue: jianlun chuanboxue zai woguo de fazhan fangxiang [Communication study is propaganda study: A discussion on the direction of the development of communication study in China]. *Zhengzhou daxue xuebao, 3*, 70–77.

Li, D. (1992). *Shiyong xuanchuan xue* [Applied propaganda science]. Jinling shushe.

Li, D., & Guo, Q. (1992). *Shiyong xuanchuan xue* [Applied propaganda science]. Zhengzhi.

Liang, Q. (1923). *Xianqin zhengzhi sixiang shi* [A history of political thought in Pre-Qin]. Shangwu yinshu guan.

Liang, S. (1936). *Shiyong xuanchuan xue* [Applied propaganda science]. Shangwu yinshu guan.

Lin, C. (2013, November 21–24). *Red is not an answer: Rethinking propaganda in the Mao era* [Presentation]. National Communication Association Conference, Washington, DC.

Lin, C. (2015). Red tourism: Rethinking propaganda as a social space. *Communication and Critical/Cultural Studies, 12*(3), 328–346. https://doi.org/10.1080/14791420.2015.1037777

Lin, C., & Nerone, J. (2015). The "great uncle of dissemination": Wilbur Schramm and communication study in China. In P. Simonson & D. W. Park (Eds.), *The international history of communication study* (pp. 396–415). Routledge.

Lindeman, E. C., & Miller, C. R. (1940). Introduction. In H. Lavine & J. A. Wechsler (Eds.), *War propaganda and the United States*. Yale University Press.

Linebarger, P. M. A. (1948). *Psychological warfare*. Infantry Journal Press.

Liu, H. (2013). *Xuanchuan: guannian, huayu jiqi zhengchanghua* [Propaganda: Concept, discourse, and legitimacy]. Zhongguo dabaike quanshu chubanshe.

Liu, J. (2011). Zhongguo gongchuandang xuanchuanjia shi chuanboxue zhuyao yuanli de shouchangzhe [The CPC propagandists were the original creators of basic principles of communication study]. *Xiandai chuanbo, 10*, 37–42.

Lu, X. (2005). *Luxun quanji (di si juan)* [Complete works of Lu Xun (vol. 4).] Renmin wenxue chubanshe.

Mann, J. (2007). *The China fantasy: Why capitalism will not bring democracy to China*. Penguin.

Mao, Z. (1991). *Mao Zedong xuanji (di wu juan)* [Selected works of Mao Zedong (vol. 5)]. Remin chubanshe.

Martin, E. D. (1929). Our invisible masters. *Forum, 81*, 142–145.

Marx, K. (1970). *Critique of Hegel's "Philosophy of Right"*. Cambridge University Press.

Marx, K. (1998). *The German ideology*. Prometheus.

Mattelart, A. (1996). *The invention of communication*. University of Minnesota Press.

McLaurin, R. D. (Ed.). (1982). *Military propaganda: Psychological warfare and operations*. Praeger.

Miller, C. (1937). *Propaganda analysis*. Institute for Propaganda Analysis.

Minkov, M., & Hofstede, G. (2011). The evolution of Hofstede's doctrine. *Cross Cultural Management: An International Journal, 18*(1), 10–20.

Nerone, J. (2015). *The media and public life: A history*. Polite.

Parry-Giles, S. J. (1994). Propaganda, effect, and the Cold War: Gauging the status of America's "war of words." *Political Communication, 11*(2), 203–213.

Pearlin, L. I., & Rosenberg, M. (1952). Propaganda techniques in institutional advertising. *Public Opinion Quarterly, 16*(1), 5–26.

Perry, E. J. (2013). *Cultural governance in contemporary China: "Re-orienting" party propaganda*. [Harvard-Yenching Institute working paper series]. http://www.harvard -yenching.org/sites/harvard-yenching.org/files/featurefiles/Elizabeth%20Perry_ Cultural%20Governance%20in%20Contemporary%20China_0.pdf

Perry, J. (1942). War propaganda for democracy. *The Public Opinion Quarterly, 6*(3), 437–443.

Pratkanis, A. R., & Aronson, E. (2001). *Age of propaganda: The everyday use and abuse of persuasion*. W. H. Freeman.

Qiu, Z. (1993). *Shijie xuanchuan jianshi* [A brief international history of propaganda]. Fujian renmin chubanshe.

Qualter, T. H. (1962). *Propaganda and psychological warfare*. Random House.

Ricoeur, P. (1986). *Lectures on ideology and utopia*. Columbia University Press.

Rogers, C. B. (1949). *The spirit of revolution in 1789*. Princeton University Press.

Rohatyn, D. (1988). Propaganda talk. In T. Govier (Ed.), *Selected issues in logic and communication* (pp. 73–92). Wadsworth.

Romerstein, H. (2009). Counterpropaganda: We can't win without it. In J. M. Waller (Ed.), *Strategic influence: Public diplomacy, counterpropaganda, and political warfare* (pp. 137–180). Institute of World Politics Press.

Sawyer, J. K. (1991). *Printed poison: Pamphlet propaganda, faction politics, and the public sphere in early seventeenth-century France*. University of California Press.

Schettler, C. H. (1950). Propaganda techniques—past and present. *Quarterly Journal of Speech, 36*(1), 78–80.

Shambaugh, D. (2007). China's propaganda system: Institutions, processes and efficacy. *The China Journal, 57*, 25–58.

Shi, X., & Wang, J. (2011). Cultural distance between China and US across GLOBE model and Hofstede model. *International Business and Management, 2*(1), 11–17.

Shi, Z., & Gao, Y. (2011). *Shiyong xinwen xuanchuan xue* [Applied science of news propaganda]. Zhongguo chuanmei daxue chubanshe.

Socolow, M. J. (2007). "News is a weapon": Domestic radio propaganda and broadcast journalism in America, 1939–1944. *American Journalism, 24*(3), 109–131.

Sproule, J. M. (1987). Propaganda studies in American social science: The rise and fall of the critical paradigm. *Quarterly Journal of Speech, 73*(1), 60–78.

Sproule, J. M. (1989a, August). *Propaganda: Five American schools of thought.* The Biennial Convention of the World Communication Association, Singapore.

Sproule, J. M. (1989b). Social responses to twentieth-century propaganda. In T. D. Smith (Ed.), *Propaganda: A pluralistic perspective* (pp. 5–22). Praeger.

Sproule, J. M. (1997). *Propaganda and democracy: The American experience of media and mass persuasion.* Cambridge University Press.

Stockmann, D. (2013). *Media commercialization and authoritarian rule in China.* Cambridge University Press.

Szunyogh, B. (1955). *Psychological warfare: An introduction to ideological propaganda and the techniques of psychological warfare.* William Frederick.

Tong, B. (2001). Kexue he renwen de xinwenguan. [The scientific and humanitarian perspectives of journalism]. *Xinwen daxue, 2*, 5–9.

Tuchman, G. (1972). Objectivity as strategic ritual: An examination of newsmen's notions of objectivity. *American Journal of Sociology, 77*(4), 660–679.

Wang, F. (n.d.). *Xuanchuan xue* [The science of propaganda]. Henan junshi zhengzhi ganbu xunlianban.

Wang, X. H. (1994). *Xuanchuan xue yinlun* [Introduction to the science of propaganda]. Hangzhou Daxue chubanshe.

Wang, X. L. (2010). Minguo shiqi gonggong guanxi jiaoyu chuangjian shimo [The beginning of the public relations education in Republican China]. *Xinwen yu chuanbo yanjiu, 6*, 55–60.

Wang, Y. (1944). *Zhonghe xuanchuan xue* [Comprehensive propaganda science]. Guomin tushu chubanshe.

Wang, Z. (1982). Lun xuanchuan [On propaganda]. *Xinwen daxue, 3*, 5–10.

Weber, M. (1946). *From Max Weber: Essays in sociology.* Oxford University Press.

Weber, M. (1958). *The protestant ethic and the spirit of capitalism.* Scribner.

Weber, M. (1964). *The religion of China.* The Free Press.

Weber, M. (2004). *The vocation lectures.* Hackett.

Whitton, J. B. (1951). Propaganda in Cold Wars. *Public Opinion Quarterly, 15*(1), 142–144.

Williams, R. (1961/1975). *The long revolution.* Greenwood.

Williams, R. (1980/1997). *Problems in materialism and culture.* Verso.

Williams, R. (2003). *Television.* Routledge.

Wong, M. A. (2002). China's direct marketing ban: A case study of China's response to capital-based social networks. *Pacific Rim Law & Policy Journal, 11*(1), 257–284.

Wu, C., & Wu, D. (Eds.). (1695/1995). *Guwen guanzhi* [A selection of classical Chinese essays]. Jiangsu wenyi chubanshe.

Zhang, J., & Cameron, G. T. (2004). The structural transformation of China's propaganda: An Ellulian perspective. *Journal of Communication Management, 8*(3), 307–321.

Zhang, K., & Qian, X. (1992). *Xuanchuan xue daolun* [Introduction to the science of propaganda]. Xueyuan chubanshe.

Zhao, Y. (1998). *Media, market and democracy in China: Between the party line and the bottom line*. University of Illinois Press.

Zheng, B. (1987). *Xuanchuan xue gailun* [Introduction to the science of propaganda]. Liaoning daxue chubanshe.

Zhuzi jicheng [Selected works of Pre-Qin thinkers]. (2006). Zhonghua shuju.

3   Tourism as a propaganda system

Like propaganda, "tourism," too, is a confusing term as it designates too many things (Rojek & Urry, 1997). However, there are generally two approaches to studying tourism: professionals regard tourism as a business and academics see it as a social phenomenon (Apostolopoulos et al., 1996). In this book, I combine the two distinct approaches, regarding tourism as a *social space*, comprising both discursive networks (e.g., history, politics, media, etc.) and non-discourse (e.g., economy). Put very simply, tourism is an amalgam of history, politics, culture, collective memories, mediated communications, economy, and more. The first step of rethinking tourism as a social space is to think of it as a system, specifically a media system, more specifically a propaganda system.

Whether referring to "travelling culture" from an anthropological perspective (Clifford, 1992; Said, 1985) or "touring culture" (Rojek & Urry, 1997) from a sociological perspective, the conceptual constructions of tourism propel the contemporary critical study of tourism into the cultural domain where media studies as a field has occupied an important seat. Nevertheless, a quick catalog of the authors who contribute greatly to what one may call "tourism studies" is dominated by sociologists and anthropologists; media scholars as a group are largely invisible. This is in part because we, media studies scholars, tend to adopt ideas and lexicon from social theorists that are not necessarily within our domain. This is not to suggest that there has been a lack of interest in studying human travel in the field of media studies. On the contrary, studying mass mobility and space is indeed a legacy bequeathed by the early generation of communication and media studies scholars, noticeably James Carey and Armand Mattelart among others. But we do not have a theoretical framework of tourism that speaks directly to/and about either media or mediation. Grabbing existing constructs from social/critical theorists without significant deconstruction and reconstruction would likely bring us into the endless circle within the realm of the sense of the Western leisure class, which is irrelevant to the current study of Red Tourism.

In this chapter, I attempt to reconstruct tourism as a propaganda system that operates through a mechanism whereby ideological messages are channeled into the market, disseminated by human travel, and (re)produced by consumption. The proposed tourism-propaganda framework (TPF) is a new scheme of tourism constructed in the field of media studies. It connects and converges existing ideas and

DOI: 10.4324/9781003231783-3

theories concerning popular culture and media, tourism, propaganda/*xuanchuan*, and the public sphere into a single frame through which we are not only able to see the tourist, the public, the state, and ideology differently against the backdrop of tourism but also the dynamics among these critical elements of the sphere of tourism beyond a phenomenology of leisure and recreation. TPF has a triad of theoretical dimensions: mass culture, mass tourism, and mass communication. It is on the basis of this triad of masses that my texts are organized. The construction of the TPF is multipurpose: it partly serves as a continuation of the previous chapter, partly a theoretical preparation for the investigation of Red Tourism in subsequent chapters, and partly an indispensable conceptual grasp with which this research is conceived, constructed, and purposed.

Popular culture as popular propaganda

There are many ways to approach tourism, such as a sociology of tourism or an ethnomethodology of sightseers for example. My theoretical point of departure, however, is popular culture, sometimes also referred to as popular media or mass culture. This is because in the final analysis, tourism is effectively a form of popular culture.

Scholars have pointed to the propagandistic nature of popular culture without using the term "propaganda." What I do herein is to connect those seminal thoughts and translate them into the thinking around propaganda. In doing so, I want to pinpoint that popular culture is a popular site for popular propaganda. This is of pivotal importance as this study no longer treats propaganda as the state's deceptive manipulation, but a produced yet productive space encompassing both the state and the private. In other words, the reinterpretation of popular culture brings the previously unspotted private sector into light in producing, distributing, and circulating propaganda. It assumes that consumers of mass culture are mobilized for producing popular propaganda while being propagandized. This, again, devotes our attention to a profound social space of propaganda and, at the same time, turns the folklore notion of propaganda on its head.

In light of the *xuanchuan* model of propaganda, popular culture can be viewed as a modern *xuanchuan* practice within which intended ideologies, for good or ill, are spread out *horizontally* through human (popular) social networking, or "social communication" in Williams' (2003) description. In this process, previously passive indoctrination becomes active consumption, manipulation camouflages as marketing, ideologies come in the guise of fictional narratives or visual-audio attractions, and finally, hegemony arises out of popularity. The upshot is that propaganda, a previously dull form of media communication, is now to be perceived as rather fantastically commercial. This is, in part, why tourism almost completely loses its ideological connotation in the public's imagination.

The propaganda nature of the culture industry

The tangent between the line of propaganda and the circle of popular culture is fear. The previous fear of propaganda is now entering the realm of popular culture.

Unsurprisingly, intense debates on popular culture have very much focused on the effects of popular culture that can roughly fall into two lines of research: strong media effect and limited media effect. This is to say, the paradigm of the study of popular culture, to a great extent, and that of propaganda studies are conflated.

However, in popular cultural studies scholars prefer a new vocabulary of propaganda. In that lexicon, for example, "cultural dupe" and "one dimensional man" replace "puppet/propagandee," and "indoctrination," "prescribed attitudes," the "Happy Consciousness," and "myth" substitute for "propaganda." I am not suggesting that this is wrong; conversely, it, again, confirms the complexity of the social space of propaganda. Rather, I want to make a point that in reality, the study of popular culture signposts a regime of propaganda where culture, ideology, politics, and economy are intertwined, reassembled, and repacked in the name of commodity. Fiske (1987) described popular culture as a battlefield of "semiotic guerrilla warfare" (p. 316). Against this way of thinking, I contend that it would be more precise to consider popular culture the battlefield of "propaganda guerrilla warfare" where all sorts of ideologies fight for both consumers and hegemonic power. Perhaps, a more positive thinking would be that popular culture is the marketplace of propaganda.

In what follows I will explore the nexus between popular culture and propaganda. This means that although I will to a certain extent sketch out the tradition of cultural studies of popular culture, this will be done only as it can be extended to shed some light on propaganda; so that I will be very selective in terms of which aspects of cultural studies I choose for discussion. Since the current study is theoretically founded on two strands of Marxism, classical Marxism and Western Marxism, I draw heavily from scholars in the Marxist tradition, namely, Theodor Adorno, Max Horkheimer, Herbert Marcuse, Walter Benjamin, and Raymond Williams. Each, in their different way, provides me with insights into the conjuncture of popular culture and propaganda. What unites them is an underlying assumption that popular culture is a social form of power full of potential for spreading ideologies.

Up to a certain point, the standard Frankfurt School critique of popular culture can be seen as the equivalent of a critique of propaganda. Storey (2009) pointed out that the Frankfurt School treats mass culture as "imposed culture of political manipulation" (p. 82). Adorno's idea of the culture industry as mass deception is a characteristic example. In *Dialectic of Enlightenment*, Horkheimer and Adorno (2002) deliberately replace "mass culture" with a coined term "the culture industry" in an attempt to discredit the idea of popular culture as a "culture that arises spontaneously from the masses themselves" (Adorno, 1975, p. 12). For Horkheimer and Adorno, the culture industry is deceptively pernicious. They contend that the idea that "they [the cultural industries] are nothing but business is used as an ideology to legitimize the trash they intentionally produce" (Horkheimer & Adorno, 2002, p. 95). In other words, for Horkheimer and Adorno, what popular culture does is to propagandize as it seeks to confirm the established social order and to maintain ideological hegemony. Characterized as the mass proliferation of sameness, homogeneousness, universality, and inflexibility in the popular culture sphere, ranging from film and radio to architecture, the culture industry is believed to be the inevitable outcome of monopolistic capitalism.

The ghost of Western propaganda also haunts the field of popular culture. Horkheimer and Adorno (2002) note, "[u]nder the private monopoly of culture tyranny ... sets to work directly on the soul" (p. 105). Adorno (1975) fully articulated the propaganda nature of popular culture in his later account of the culture industry: "[t]he total effect of the culture industry is one of anti-enlightenment, in which ... enlightenment, that is the progressive technical domination of nature, becomes mass deception and is turned into a means for fettering consciousness" (pp. 18–19). It is not an exaggeration that Horkheimer and Adorno's idea of the culture industry can be interpreted as a transmutation of Herman and Chomsky's propaganda model in popular culture. In both cases, text and practice of popular culture, whether in the form of hit songs, films, or news coverage, ultimately become propaganda through filters of capitalism and the ruling ideology.

One-dimensional man is believed to be the final product of the culture industry. Marcuse (2002) viewed popular culture as a means of social control through which man and his behavior and thought become one dimensional, ultimately resulting in a one-dimensional society alongside a one-dimensional reality. This is in part because, as Marcuse explained, compared to what he called "the pre-technological culture" (high culture), the dimension of truth is lost in mass culture. Specifically, mass reproduction to Marcuse is the process through which the logic of Reason turns into the logic of domination. In other words, it is on this basis that popular culture becomes propagandistic. Propaganda is delivered to the consumer as "prescribed attitudes and habits, certain intellectual and emotional reactions" (Marcuse, 2002, p. 14). These are propagandas because they "indoctrinate and manipulate; they promote a false consciousness which is immune against its falsehood" (Marcuse, 2002, p. 14). For Marcuse, the ostensible democratization of popular culture is merely the domestication of culture insofar as the antagonistic power of artistic alienation is removed to preserve social domination. Marcuse (2002) contended:

> This sort of well-being, the productive superstructure over the unhappy base of society, permeates the "media" which mediate between the masters and their dependents. Its publicity agents shape the universe of communication in which the one dimensional behavior expresses itself. Its language testifies to identification and unification, to the systematic promotion of positive thinking and doing, to the concerted attack on transcendent, critical notions.
>
> (p. 88)

It is worth noting further that the critique of popular culture by classical Marxists is oftentimes misunderstood, generating a substantial number of counterarguments to the culture industry. In fact, what the classical Marxists assumed was not a completely passive audience incapable of critical thinking but a kind of technology-and-mass-consumption-camouflaged superpower effective in implanting ideologies into the masses. Whether the audience is intelligent or unintelligent is irrelevant to their critiques, because the capitalist mode of production determines the nature of the culture industry regardless of the consumer's agency. If Adorno, Horkheimer, and Marcuse advocated a strong-effects model of mass culture, other Marxist theorists such as Walter Benjamin and Raymond Williams can be seen as proponents of a limited-effects model. They envisaged a more active mass in the site of consumption.

Referring to mass culture as technological reproduction, however, Benjamin's (2008) account did not focus on technology; he spoke volumes about ideology instead. For Benjamin, the technological is essentially ideological and the result of technological reproduction is an organized way of seeing. Accordingly, Benjamin's idea of popular culture has two fundamental parts: politics at the site of production and interpretation at the site of consumption. For Benjamin (2008), popular culture represents a paradoxical situation that he referred to as a "shattering of tradition": on the one hand, he explained, it is "destructive, cathartic," and on the other hand, it means the "liquidation of the value of tradition in the cultural heritage" (p. 22). Through the circle of mass production, mass propagation, and mass perception, diverse ways of seeing were increasingly being organized into the prevailing way of seeing, as Benjamin (2008) suggested, "the way each single image is understood seems prescribed by the sequence of all the preceding images" (p. 27).

Benjamin did see some degree of control on the audience end but he saw this rather dialectically. Benjamin (2008) argued that: "[i]t is they [the masses] who will control him [the actor]. Those who are not visible, not present while he executes his performance, are precisely the ones who will control it. This invisibility heightens the authority of their control" (p. 33). Benjamin goes further reminding us that the audience's control has no political power as long as the capitalist model of production is still on the scene. Contrarily, Benjamin indicated that the masses' revolutionary control over cultural products can be repurposed for counterrevolutionary use under capitalist circumstances. Another control is about literacy. In the wake of mass production, literacy competence now becomes the common property of the masses; but the other side of the coin is that the capitalist also takes advantage of it to "distort and corrupt" class consciousness of the masses (Benjamin, 2008, p. 34).

Labeled as "culturalism" by other scholars, Williams' grasp of mass culture is more nuanced. For Williams (1961/1975; 1989), culture by and large is a particular way of life, and he repeatedly reminded us that culture is ordinary (Williams, 1989). In Williams' framework, culture has three levels: the lived culture, the recorded culture, and the selective tradition. The selective tradition is of crucial importance among the three for it not only links the recorded and the lived but also produces them. Williams (1961/1975) has argued that the process of selection and re-selection is governed by the "contemporary system of interests and values," in a nutshell, the prevailing ideology (p. 68). But what distinguishes Williams from other Marxist scholars is that popular culture for Williams is not merely a matter of selection on the producer end; it is also a matter of selection on the audience end. This notion of double-selection is consistent with Williams's (1960) conviction that "[c]ommunication is not only transmission; it is also reception and response" (p. 332). It resembles Lefebvre's idea of space to the extent that symbolic meaning is both produced in the site of production and (re)producing new meanings through interpretation. In the same vein, Williams (1989) rejected the idea of the masses as the duped, arguing that "there are in fact no masses, but only ways of seeing people as masses" (p. 11). Exemplifying the reading habit of his father, Williams (1989) made an important point that high quality of intelligent life does not prevent one from enjoying popular culture. In other words, Williams refuted a general presumption that the proliferation of mass culture leads to the decrease of public intelligence. In his study of television, Williams (2003) viewed the effects of popular culture as social changes (e.g., his notion of mobile privatization) within larger social changes. In other words, William rejected the cultural degradation thesis dominant in the Frankfurt School's critique of popular culture.

There is a popular myth bearing the coded and alternative vocabulary of propaganda in popular culture studies. The myth goes like this: upon propaganda entering the realm of popular culture, it is no longer as propagandistic as it is in the state sector. This mode of thinking has shifted the focus of our attention away from hegemony to the ideology of consumption, thus reducing a wide range of ideological mediations to a simple matter of commercialism. But I argue otherwise. Specifically, I argue that propaganda becomes even more powerful after being reproduced and repacked as "fantasy" in various popular cultural forms than being produced by state propaganda organs. Along the same line, I contend that popular culture is *the* popular site of propaganda.

This proposition can be understood in two ways. First and foremost, propaganda in this terrain is reproduced in popular forms of communication such as film, pop song, talk show, etc. In other words, propaganda can be popular since it no longer comes in dull text but something aesthetically or sensually appealing. Second, popularity also indicates the lively and heavy traffic in exchanging propaganda in the marketplace. That is, as the *xuanchuan* model suggests, propaganda now becomes a case of everyone propagandizing everyone else through social communication, hence popular propaganda.

In the preceding chapter, I have remarked that the most popular and persistent propaganda in China has been the *zhong-xiao* (loyalty-filial piety) propaganda. Here I turn to racism, arguably the most popular propaganda in the modern West. A powerful propaganda, racism was crucial in propagating colonialism and conquering colonial nations worldwide. It sugar-coated colonial conquest "as if directed by God" (Storey, 2009, p. 171). Disseminated through popular culture, this popular propaganda back in the 19th century assumed that the Negro, inferior to Western Europeans, could only be civilized and humanized by White people. Justified by such racist propaganda, colonialism and imperialism were masked as a civilizing mission for the well-being of the conquered (Fryer, 1984). Yet the topic of racism has been rarely talked about in the language of propaganda. Among many forms of racism, Orientalism is an exotic and popular one.

Said (1977) observed a propaganda "layer" of Orientalism manifested in the institutionalization of studies of Oriental languages in the West. It would have been more illuminating if Said could say more assertively that Orientalism is propaganda, rather than referring to it as a "system of ideological fiction" (p. 321). The propaganda feature becomes plainly evident in that "[i]t [Orientalism] is one of the mechanisms by which the West maintained its hegemony over the Orient" (Said, 1977, p. 300). This is to say, like propaganda, the ultimate goal of Orientalism is to gain hegemonic power; but unlike pure political propaganda, the territory of Orientalism is popular culture where Orientalism as propaganda is produced in diverse forms of fantasy in the private, narratively penetrating and aesthetically alluring.

Many reasons can explain why people appear reluctant to refer to Orientalism as propaganda and at least two of those are worthy of special mention. First, there is no political party, political institution, government, political leader, or state that would take the responsibility for initiating such propaganda; there are myriad instead: various countries, various countries' political institutions, various countries' cultural industries in addition to millions of professionals. Said (1977) simply deemed it as "almost European invention" (p. 1). This does not fit the existing popular imagination of propaganda as the province of certain countries, governments, political parties, and political leaders and authorities. However, no scholars have said that propaganda needs to be that way. Rather, it is an imagination based on many other imaginations and the Representation (re)produced by media representations. The Orientalism case illuminates how popular propaganda is produced and producing in the private sector. Second, the spectacular makeover after being commodified renders the propaganda quality of Orientalism so imperceptible. The exotic feature of cultural products of Orientalism has been perceived as "edge" and "chic," sometimes even classic. Consider Hollywood's *Shanghai Express* (1932), *The Good Earth* (1937), *Casablanca* (1942), *The World of Suzie Wong* (1960), *Cleopatra* (1963), among many others. To the extent that their narratives seek to legitimate, indoctrinate, and propagate intended ideologies, these films are propagandistically resonant. During the Cold War, the United States produced and consumed a vast volume of fantasies set in Asia. Christina Klein (2003) has pointed out that the growing interest of the US film industry in Asia during the late 1940s and 1950s was closely related to the political, military, and economic expansions of the United States in that region. In 1950, Raymond A. Hare, acting assistant secretary of state for Near Eastern, South Asian, and African Affairs spoke to the Under Secretary of State in a meeting:

> What we have to do is to convince not only their minds but their hearts. What we need to do is to make the "cold war" a "warm war" by infusing into it ideological principles to give it meaning.
>
> (quoted in Klein, 2003, p. 19)

I do not think it is too much to say that from colonial conquest to the Cold War, the global expansion of Western powers depended considerably on the deployment of popular propaganda in diverse popular cultural forms. It should be noted further that Orientalism has been always bound to space, which Said (1985) refers to as "imaginative geography" characterized by an invisible yet defined line "separating Occident from Orient" (p. 90). As such, tourism is a popular site of Orientalism. John Urry would see Orientalism as the Western gaze on the Oriental. This Orientalist gaze is evident in the contemporary tourist consumption of belly dance in Istanbul (Potuoglu-Cook, 2006). In fact, it has become a selling point for promoting indigenous tourism by many Asian countries such as China (Yan & Santos, 2009) and Oman (Feighery, 2012).

My claim that popular culture is propagandistic, however, does not extend to say that popular culture exists merely to manipulate people's minds. With its neutral connotation as in the *xuanchuan* model, propaganda in the context of popular culture cuts in two ways: propaganda as domination and propaganda as resistance. In other words, popular propaganda in the private sector works in two directions: one towards reinforcing the established social order and power and the other one towards challenging such hierarchy, as Foucault (1978) famously noted, "Where there is power there is resistance" (p. 95). Popular discourse of the Vietnam War in US history is a powerful case in point.

Two kinds of popular propaganda were noticeable in American popular culture during the Vietnam War. The one that sold the war to American audiences is characterized by Hollywood's Vietnam. As a genre, the Vietnam War film produced a "particular regime of truth" (Storey, 2009, p. 176). It is a transformation of state propaganda into popular discourse/propaganda as it narratively confirms Americanization/Americaness in a rather narcissistic way that the war in the diegesis is not about Vietnam and the Vietnamese, but all about the Americans. As a result, as Klein (1990) suggested, "the war is decontextualized, mystified as a tragic mistake, an existential adventure, or a rite of passage through which the White American Hero discovers his identity" (p. 10). In contrast, the other kind of propaganda opposes the war. It was embodied in protest music of the sixties such as Bob Dylan's "Masters of War" (1963), Phil Ochs' "Talkin' Vietnam Blues" (1964), Neil Young's "Ohio" (1970), and so forth. Anti-Vietnam protest music coexisted with folk music concerning Civil Rights and nuclear disarmament in the 1960s. Auslander (1981) argued that the shift of rock and folk music from these movements to anti-Vietnam War was anticipated by the escalation of the American presence in Vietnam. Then, can popular culture be seen as the free market of propaganda? While it is a market with competing ideas, ideologies, and propagandas, it is by no means "free." Here the capitalist mode of production matters. In content analyzing American popular music surrounding the representation of American soldier between 1965 and 1985, James (1989) demonstrated that the dissident voice was only marginal in contrast to the hegemonic voice from the culture industry. In other words, the two types of propaganda were imbalanced with regard to media representation.

The triumvirate-sphere of popular culture

The problem of the culture industry is the problem of propaganda. The great divides between strong-effects and limited-effects proponents and culture industry and human agency advocates stem from and manifest a substantial divergence in understanding Marx's view of human society incarnated in the dialectical relationship between the base and the superstructure. I agree with Williams' and Lefebvre's interpretations. Both authors, though approaching it differently, reject a mechanical reading of the relationship as a combination of the determining base and the determined superstructure. Williams (1980/1997) sees a mediation between the dynamic base and superstructure, whereas Lefebvre (1991) proposes a social space both produced and productive. I will elaborate their points later in a chapter on "The Social Space of Red Tourism."

My take on the debate over the effects of mass media culture is twofold. On the one hand, I do not believe agency, understood as individual capacity, can counteract the power of structure. The most imbalanced US news coverage on the Gulf War debate is a straightforward example in this regard (Fico & Cote, 1999). On the other hand, I do not deny the power of the general public either. Nevertheless, I would argue that to be able to challenge structural power, individuals must work collectively as/in a structure. For example, I do not think of an anti-racism movement such as Black Lives Matter as an instance of individual agency fighting against structure. Rather, I believe this is an instance of one structure challenging another; but this is inconceivable in the agency–structure binary. Here I need a new model.

My solution to the problem of mass culture is a triumvirate-sphere of popular culture consisting of propaganda, the culture industry, and the public sphere (Figure 3.1). It postulates that in the space of popular culture there exists the triad of spheres: the propaganda sphere, the culture industry, and the public sphere. It is a deconstructing and remobilizing of the mentality of the Frankfurt School based on the *xuanchuan* model. Specifically, it is a reassembling of Horkheimer

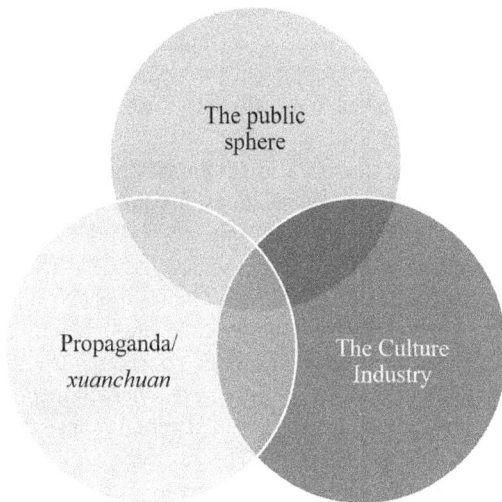

The public
sphere

Propaganda/
xuanchuan

The Culture
Industry

Figure 3.1 The triumvirate-schema of popular culture. Created by the author.

and Adorno's idea of the culture industry, Habermas's public sphere, and the cultural practice of *xuanchuan*. The dynamics of the triumvirate sphere determine the nature of Red Tourism, which cannot be understood as a domain of state propaganda but a social space where the forces of propaganda, economy, and public opinion are inextricably intertwined, together producing a social imaginary of contemporary China. I have discussed *xuanchuan* and the culture industry, now I turn to the public sphere.

Note that there is an idea of the "public" and there is also one of the "mass." The public practices democracy in the public sphere, but what do the masses do? And where? Is there a "mass sphere"? To answer these questions, the pejorative connotation of mass/masses might be a starting point.

Assuming mob-status, the term of "masses" signifies "gullibility, fickleness, herd-prejudice, lowness of taste and habit" (Williams, 1960, p. 317). According to Williams, the meaning of masses is threefold. First, it means physical massing, designating the human movement from the rural to the urban, hence urbanization. Second, the massing became social when workers were organized into factories, ultimately leading to mass production. Third, the development of class consciousness of the working class rendered the massing political, resulting in mass action. Characterized as "mass-thinking," "mass-suggestion," and "mass-prejudice," the masses were believed to threaten what Williams referred to as "class-democracy." By class, he really meant the ruling class. No doubt the gullible working-class masses in this context were set up against the intelligent middle-class reading public, but a question inevitably arises here with Habermas' construct of the public sphere: were the masses as a group included in the public sphere? Surely, the working people were not the regulars of coffee shops, salons, who were described as major actors of the bourgeois public sphere. If the masses, the majority, were excluded from the public sphere, then where did they practice democracy? These are crucial questions that I cannot sidestep. Given the fact that China is not a Western democratic country, conjuring up a Chinese public sphere has been extremely difficult, if not utterly impossible. But Chinese citizens must have a "sphere" in which critical issues are brought out, discussed, and debated in the public domain. To tackle this requires much imagination.

The theory of public sphere has gained currency in China starting from the 1980s. The heated debates have centered on the interrelationships among a series of concepts, mainly, "state", "civil society," and "private sphere," and their applicability to Chinese society (Xia & Huang, 2008). No consensus has been made as to whether there is a counterpart of the public sphere in China, if it exists at all, and what it should be. Nevertheless, Chinese scholars hold the unanimous conviction that public sphere theory has to be reworked to be considered useful in China. While various attempts to "transplant" this Western democracy theory to a non-Western-democratic state have demonstrated the intellectual robustness of Chinese scholarship, the result is rather disappointing. Theoretically, a Chinese public sphere is still unimaginable. The difficulties are discernable.

A thorny issue is about vocabulary. The original vocabulary of the Habermasian public sphere works well in early modern Europe. This is in part because the theory itself was developed on the discourse of German bourgeois society (Habermas, 1989). But this discourse alongside its practice is alien to the Chinese history and culture. Even the Chinese translation of the term "public sphere," or *gongong lingyu* 公共领域 (literally, "public domain") sounds administratively familiar but can be easily misleading. Secondly, the debates on the aforementioned core concepts without referring to specific historical accounts have made this line of Chinese scholarship even more abstract than the original work of Habermas. Thirdly, the major players and places of the Habermasian public sphere are also foreign to Chinese culture. Merchants, bankers, manufacturers, and entrepreneurs, perhaps, are the last group of people in the minds of Chinese people when conjuring up an image of the general public. Moreover, the places in the Habermasian public sphere such as coffee houses and salons are equivalently foreign. All of these problems suggest that a new model is in need. Here the cultural practice of *xuanchuan* may help conceive of the Chinese public sphere.

As noted earlier, *xuanchuan* is meant to empower the people and aims for the good of society with close ties to education. This benign feature points to an alternative public sphere. The big problem, however, is that the primary function of *xuanchuan* lies in its governance, whereas the public sphere, whatever it is, is headed for the opposite, democracy.

A possible way to solve this puzzle is to relook at the private sector. *Xuanchuan* can be initiated from the private. Historically, such *xuanchuan* /propaganda from below has led to multiple revolutions and democratic movements in China. As a matter of fact, democratic movements in modern China, say, the May Fourth Movement of 1919, have been considered to be sparked from *xuanchuan*. As noted in the preceding chapter, some American scholars hold a similar view that even propaganda in the West can work for democracy. Taking this as a point of departure, I argue that the cultural practice of *xuanchuan*, in particular, can help construct an alternative public sphere by inviting us to rethink "public," "sphere," and "media," the three fundamental elements of the public sphere framework in the Chinese context.

First of all, the mobility in the *xuanchuan* model points to a mobile public. Culturally, Chinese people have long imagined the nation-state as a big family (Sun, 2000). In this cultural imagination, the boundary between the state and the private is believed to be difficult to be drawn (e.g., Huang, 1993; Liang, 2003). To avoid the trap of an already abundant rhetoric about what should be and what should not be included in the public sphere in Western scholarship (e.g., "gendered publicity," "counterpublics," and "subaltern counterpublics") and particularly since my effort here is only to invoke an image of a Chinese public, I will not engage in this long line of research, which I think merits a separate project. My point is that every citizen should be included in the mobile public.

Related is a mobile sphere. It does not require a set of fixed places. The word-of-mouth method of the *xuanchuan* practice suggests that propaganda can be made through social communication. There are no designated places for this kind of communication; it can happen anywhere and everywhere. This is also the case for the proposed Chinese model of the public sphere. That is to say, public debates on critical issues in China can be sparked off in any kind of space, physical, virtual, and/or abstract spaces. Roughly, the mobile sphere includes four types of space: (1) public space such as the passenger cars of trains, restaurants, classrooms, and workplaces. Lord Bryce, for example, considered smoking cars of commuter trains the public space where public opinions were exchanged (DeFleur, 1998; Katz, 2006); (2) private space such as family; (3) virtual space like micro-blogs (Weibo and WeChat), online forums, and so forth; (4) abstract space such as regional operas (*difang xi* 地方戏) and many other types of folk arts.

   Family is a characteristic example. People in the West do not normally think it is appropriate to include what Habermas refers to as the intimate sphere into the public sphere because "public" and "intimate" are mutually exclusive within a dualistic model. Nevertheless, family plays a pivotal role in political communication. Lazarsfeld has shown in the famous presidential election research (aka the Decatur studies) that compared to news media, family members, and close friends were no less important in their political decisions (Lazarsfeld et al., 1944; Katz, 2006). This is where Lazarsfeld's account of personal influence and the "two-step flow of communication" begin. From the mobile perspective, the intimate sphere can be readily transformed into a public sphere. In a traditional society, it is common that Chinese people want to invite and to be invited by friends, relatives, and colleagues to their homes. This format of the exchange of visit is not for partying though. Socializing, maybe, but not in the format of scattered chatting and small group discussion. Although the form of home gathering in China is quite fluid, ranging from dinning-table eating, tea drinking to simply chatting, all participants are supposed to take part in the mutual talk. "All people" might be misleading here, since women and children were largely excluded in the traditional Chinese society, particularly in rural communities' pre-economic reform. While this, to some extent, demonstrates gender inequality, it is also a sign of critical discussion as true in the Habermasian public sphere. In other words, critical issues that matter to the nation-state are typical in Chinese family gatherings. When I was a kid, those issues used to be governmental policies, international relations, economic reforms, and the like. Nowadays critical debates arise from such social gatherings usually involving the Party's anti-corruption campaign and changes of leadership at all levels of government, among others. However, I am certainly not saying that all family gathering events automatically involve Habermasian-style debate as much as not everyone going to a coffee shop or salon must enjoy a critical talk in Habermas' own account. Nor am I suggesting that home gatherings in China are necessarily significant to political matters. Rather, I want to make a point that in a mobile model of public sphere, even the intimate sphere can be publicized, turning into a sort of public sphere.

   The role of mass media is prominent in the Habermasian public sphere. The mass media had helped form a reading public, which Habermas (1989) considered "the real carrier of the public" (p. 23). This is not immediately applicable to China's case. The *xuanchuan* model suggests that both word-of-mouth and mass media matter to propaganda. It means that the Chinese public sphere is not entirely dependent on mass media.

Although I argue that the public is mobile and the sphere is fluid, I advocate neither a "phantom" Chinese public nor another theory of no-public-sphere-at-all. On the contrary, I believe even in a political system that is not a Western democracy, a "sphere" through which the general public engages in deliberative democracy does exist in an alternative form. What I object along with many other Chinese scholars is duplicating the Habermasian model verbatim. In search of the Chinese public sphere, I stress the mobility of the public and space in light of the cultural practice of *xuanchuan*. To differentiate this proposed Chinese public from the bourgeois public sphere, I herein refer to it as the *xuanchuan*-public sphere. It resonates with Tarde's (2010) idea of "public space," or a "more casual" kind of public sphere in Katz's (2006) description.

The *xuanchuan*-public sphere is based on the *xuanchuan* model. In that sense, it represents a structural transformation: *xuanchuan* now is seen as the proliferation of critical debates and democratic talk through social communication in the practice of everyday life. It may be considered "counter-propaganda" in a Western sense. It assumes that anyone can be included in "the public" as long as they engage in diverse *xuanchuan* activities. Not initiated by the reading public, *xuanchuan* may take diverse forms and formats, for example, a variety of folk arts, particularly those more or less dependent on the oral than the textual. Within the mobile sphere, the previously well-demarcated and fixed lines between the state, the public, and the private, and between the civic and the intimate are blurred, collapsed, becoming fluid and transformable. In this regard, the *xuanchuan*-public sphere is a public sphere in motion. As an attempt to problematize the clear-cut boundaries between popular culture, propaganda, and public practicing of democracy, the *xuanchuan*-public sphere is intended to be more evocative than decisive. Preliminary as it is, the proposed model, with its normative power, offers alternative angles to look at the pattern and rationale of Chinese communication. This is of paramount importance in understanding the contradictory role of Red tourists in terms of disseminating the grand narrative and at the same time, reconstructing a counter-narrative in my later account.

Reframing mass tourism: towards a tourism-propaganda framework (TPF)

There is mass culture and there is also mass tourism. Nonetheless, "mass" in the tourism setting no longer connotes the mass production of homogeneous goods; rather, it denotes the producing of standardized tourists by constantly organizing their way of seeing, or the "tourist gaze" in Urry's (2002) description. The connection between mass culture and mass tourism is noticeable. Mass tourism, too, is academically set up against the backdrop of a kind of modernity (Aramberri, 2010; Cohen, 1972; 1995; Rojek, 2000) and the democratization of the elite culture in the wake of industrialization or de-elitization of tourism in particular (Boorstin, 1992; Pearce, 1982). In the grand narrative, the upshot of mass tourism, like that of mass culture, was the gullible tourist. It has been said that their visions can be manipulated into "a closed self-perpetuating system of illusions" by professionals such as media practitioners, travel agents, tour guides, and the like (Urry, 2002, p. 7). The notion of touristic dupes (just like cultural dupes) has been much debated, however, whether the tourist is gullible or sophisticated is irrelevant to my account of tourism as a system of propaganda simply because my focus here is what tourism does, not what the tourist does. On the one hand, just because there is a propaganda system it does not follow that tourists would ultimately become "victims" of such propaganda. On the other hand, smart tourists do not prevent the presence of the propaganda system either. This is to say, treating tourism as a propaganda system, I am not indicating any degree of determinism either on the tourist end or on the system end. This is consistent with my disposition to see propaganda as a neutrally social space.

Distinct from other forms of popular culture, mass tourism involves mass human movement, "representing one of the largest peacetime movements of people, goods, services and money in human history" (Greenwood, 1972, p. 81). A related attribute is that tourism is a spatial practice, touring across space. Providing popular culture is the popular site of propaganda, mass mobility and space as two markers of tourism point to a mass communication system that is somewhat parallel to Propaganda de Fide, the archetypical propaganda network in the West. This leads me to argue that modern mass tourism can be better understood as a propaganda system from a media studies perspective. I refer to it as the tourism-propaganda framework (TPF). TPF somehow is analogous to the missionary system of the catholic church. Just as catholic missionaries propagandized the words of God through traveling, so tourists propagate all sorts of ideologies nationally and internationally through domestic and international tours. This is partly because the vast flow of tourists means the circulation of ideas carried by the tourists (Appadurai, 1996) and partly because touristic flows are subject to prevailing ideologies. A new framework of tourism, the TPF is meant to pull tourism out of the realm of the sensuous, relocating it in the realm of Representation, or that of ideology.

Revisiting tourism studies: the three filters

Existing social and cultural theories of tourism have been largely developed through abstraction of tourist motivations, experiences, and senses, not the essence/totality of tourism itself, or the system of tourism. In that sense, what some refer to as the sociology of tourism looks rather like the sociology of tourists. As a ramification of the tourist-orientation, the study of tourism has placed too much emphasis on varied experiences of tourists, losing the big picture of tourism as an amalgam of mass culture, mass mobility, and mass communication in shaping community and society. Therefore, revisiting the bulk of academic literature on tourism for the purpose of studying mass communication at the macro level requires a careful plan.

Inspired by Herman and Chomsky's (1988/2008) propaganda model, my strategy for developing the TPF is a similar one: to illustrate how modern tourism becomes a propaganda system. I focus on "filters," or some characteristic mechanisms by which Representation is disseminated through the tourism network. In other words, my conceptual grasp of tourism is organized and tied by the idea of filter/filtering. In reviewing the principal concepts of tourism studies, I identified three filters: organized tourist gazes, staged authenticity, and determined routes. This is by no means an exhaustive list as it only sketches out the systematic "filters" of tourism operating for legitimating and propagating the Representation. There are many other mechanisms and apparatuses of crucial importance in this regard such as museuming. The three filters are chosen only because these have been frequently engaged and vigorously debated by scholars. As such, the three-filter scheme seeks not to describe the tourism system, but to provoke, calling for a reconstruction of tourism at the site of ideology.

Organized tourist gazes

Hardly can any attempt of mapping out tourism literature evade John Urry's seminal idea of the tourist gaze. Derived from Foucault's (1975) medical gaze, the tourist gaze is conceptualized as an organized, systematized, and institutionalized way of seeing (Urry, 2002). In the tourist gaze, pleasure, nostalgic memory, sensuality, historical sense, and aesthetic taste that are mediated by representations (e.g., travel books, tourism advertisements, TV programs, fictions) and (re)shaped by discourses (e.g., education, history, politics) are reproduced in the touristic site, and about the same time producing new meanings (Urry, 1992). In short, the tourist gaze is a site of ideology. In an interview, Urry further articulated the ideological nature of the gaze, saying, "there are all sorts of dominant ideologies which surround travel and tourism and these dominant ideologies presuppose various kinds of visual gazes" (Franklin, 2001, p. 121). MacCannell (2001) has also remarked that the construction of the tourist gaze is "fully ideological" (p. 35). However, ideology is not what Urry's narrative is about; visualism is his analytical focus.

Had Urry paid more attention to institutional power behind visual modalities and moved beyond semiotics to the realm of ideology, the analytical framework of the tourist gaze would have been more powerful. In original accounts of the Foucauldian gaze, new institutions alongside the emergent institutional power are of paramount importance for producing, legitimating, and propagating those gazes. For example, the development of the asylum advanced the gaze of madness (Foucault, 1988) and likewise, the advent of the clinic and the formation of the medical gaze (Foucault, 1975), the birth of prison and the panoptic gaze (Foucault, 1977). Unfortunately, such analysis of power, institutions, and their dynamics has been largely missing in Urry's work.

The tourist-centered approach of tourism studies inevitably includes a chapter on human agency. Among many criticisms surrounding the tourist gaze, the most influential one is MacCannell's (2001) "second gaze," which depicts a free and critical tourist, hence a critical gaze. MacCannell reads a great deal of determinism in what he refers to as the "Urry Gaze," ranging from the tourist's motivation, site/sight choice to sightseeing and interpretation. For MacCannell (2001), the first (Urry) gaze, though popular and deceptive, is superficial and can be readily seen through by tourists, who will, then, cast the second gaze, a critical one upon the unseen. In the same spirit, Cloke and Perkins (1998) argued that tourists are able to counter the effects of tourism propaganda by using propaganda material in a creative and unintended way. In contrast to the agency-proponents, Dash and Cater (2015) emphasized the inaccessibility of the Real by tourists, suggesting that even MacCannell's second gaze is hopeless. In response, Urry, too, rejected the singularity of the tourist gaze, advocating a multiplicity (Franklin, 2001). What the Urry–MacCannell debate informs the current study is the ideologically manipulative nature of the tourist gaze upon which both authors would agree. The notion of the tourist gaze points to a potential hidden space of propaganda. The substantial divergence of their debate is rather about how powerful the organizing power of the tourist gaze is and whether tourists are gullible. This reminds us of the famous and long-lasting strong-or-weak-effect debate and a wholesale of human agency surrounding popular media and media effects in the field of communication study.

The tourist gaze proved to be political in some forms of tourism. In studying Vietnam's border tourism designed for Chinese tourists, for example, Chan (2006) argued that the discourse of the Vietnam–China relations (re)produces and authorizes both the tourist gaze and the host gaze. And propaganda surrounding the historical tensions of the two nations occupies an important position in that political discourse creates the dynamics in the tourist–host contact manifested in the "ferocious" Vietnamese host gaze. In a study of Israeli backpackers in India, Maoz (2005) revealed a somehow Orientalistic Israeli gaze which sees local Indians as feminine, primitive, vulnerable, and dirty. Maoz linked the Israeli gaze to post-imperialism and post-colonialism, in passing, by a note that the Israeli tourists colonist-wise created their own enclaves in India with name like "*hof Tel Aviv* (Tel Aviv beach)."

Staged authenticity

Equally influential in the literature of critical tourism studies is MacCannell's notion of "staged authenticity." Unlike Urry's sensuous orientation, MacCannell focuses on the tourist's motivation and his master metaphor is pilgrimage. MacCannell (1973; 1999) assumed that the modern tourist is a pilgrim-like character, who is motivated to travel by a desire for authentic experiences just as a pilgrim pursues the ultimate truth. To cater to the needs of tourists, the tourism industry seeks to turn a touristic site into sacred space through a multistage of what MacCannell terms as "site sacralization."[1] The analytical structure of staged authenticity is twofold: a need and a strategy.

The need, or what makes tourists leave home for touristic attractions, has been central to the study of tourism. Early studies have seen tourists seeking a radical break from everyday life and mundane things, whether driven by a kind of media-instilled "daydreaming and fantasy" (Urry, 2002), a "desire for contrast and escape" (Rojek & Urry, 1997), or an instinct longing for "foreignness"

(Kracauer, 1995). Like MacCannell, others who have also articulated the nexus between pilgrimage and tourism include Clifford (1992), Cohen (1988), Graburn (1977), and Shields (1990). What really differentiates MacCannell's concept of staged authenticity from others is not the conceptualization of the tourist need, but the strategy embodied in the attributive adjective "staged." It involves a series of deliberate manipulations of space, touristic experience, and representations.

First and foremost, "staged" implies a spatial manipulation. MacCannell considered tourism a social space. Unlike Lefebvre, MacCannell used the term "social space" in a broad way, simply referring to it as the social function of space. For MacCannell (1973), staging in tourism is specifically about manipulation of a particular set of space, namely, the front region and the back region. Typically, such manipulation involves "a series of front regions decorated to appear as back regions, and back regions set up to accommodate outsiders" (MacCannell, 1973, p. 602). Note that MacCannell's idea is derived from Erving Goffman's theory about social performance and space. It is Goffman (1956) who dichotomizes social space into "front" and "back." Both regions are institutionalized as each designates socially pre-established routines for certain social roles. For example, "performers appear in the front and back regions; the audience appears only in the front region; and the outsiders are excluded from both regions" (Goffman, 1956, p. 90). Also note that Goffman's theory involves two regions and three types of social actors. The possible permutations of the five manipulatable elements make Goffman's theory illuminatingly playful. For instance, a false social front for the audience, or an inauthentic back region for the outsiders. Both cases point to a lie. This is where MacCannell's (1973) "arrangements of social space" and "staged authenticity" came from.

Having situated Goffman's front-back schema in the tourism setting, MacCannell (1973) then pointed out that the manipulation of social space is indeed aimed at manipulating touristic experience, saying that:

> A mere experience may be mystified, but a touristic experience is always mystified, and the lie contained in the touristic experience, moreover, presents itself as a truthful revelation, as the vehicle that carries the onlooker behind false fronts into reality. The idea here is that a false back is more insidious and dangerous than a false front, or an inauthentic demystification of social life is not merely a lie but a superlie, the kind that drips with sincerity.
>
> (p. 599)

Eerily reminiscent of the Western notion of propaganda, "superlie" suggests an insincere manipulation of representations. MacCannell (1973) also noted that tourists have a marked tendency to enter the back region for authentic experience but what they see is rather staged, faked representations of social reality.

MacCannell's staged authenticity is especially helpful in understanding the propaganda potential of tourism. For example, guided tours to the abandoned testing site of nuclear weapons in Qinghai, a remote province of northwestern China, has lately gained popularity, resulting in the emergence of a tourist city called

"City of the Atom." The formerly prohibited secret military facility that witnessed the births of the first Chinese atomic and hydrogen bombs is now open to domestic tourists but still closed to foreign visitors. The tour, supposedly taking tourists to the "back" region, features staged relics in addition to the newly built memorial and monument. The site has an immense symbolic value for it embodies the "Two Bombs and One Satellite Spirit."

MacCannell remarked that a tour designed to invite tourists to the back region is educational since it promises to provide outsiders with insiders' insight and experience. Since the back region open to tourists is typically *staged* with "an aura of superficiality," or inauthenticity which is often perceived as authenticity, it actually directs tourists to a propaganda space that MacCannell (1973) referred to as "a new kind of social space" (p. 595). This is, perhaps, why in my interviews with officials from Chinese tourism authorities, many have repeatedly pointed out that Red Tourism is not unique to China; they see staged touristic sites for propaganda purposes everywhere in the world. One official said that the White House tour in the United States is a good example of Red Tourism because he believes that the tour promotes and propagandizes American values (interview, June 4, 2015).

Manipulating representation is both the means and result of manipulations of space and tourist experiences. In examining an alienated tourism in Paris, MacCannell (1999) argued that some very unattractive places in Paris such as the morgue, the sewers, and the stock exchange were transformed into tourist attractions to satisfy some kind of "alienated" needs of the tourists by manipulations of representations. MacCannell (1999) regarded those sites as the "concrete material representations of our most important institutions" (p. 57). The representations after the makeover secrete a mediated modernity that communicates to the tourists powerfully. An accumulative effect of such manipulations in tourism may lead to a change in tourists' worldview. For example, MacCannell (1999) has remarked a significant change of modern tourists, who transcend nature as "a force opposing man" to "something we must try to preserve" (p. 81).

Lately, the United States has witnessed a massive wave of statue removals. Many statues with historical significance were toppled, defaced, or removed because of their manipulative nature in representations, or propagandistic nature. By September 21, 2020, more than 130 Confederate monuments and other historic statues were taken down in the United States (These Confederate statues, 2020). Deemed as "unacceptable," a prominent Christopher Columbus statue at Marconi Plaza in South Philadelphia was called forth removal because it allegedly represents "false history, genocide, racism, oppression and similar themes" (Chinchilla, 2020).

Determined routes

In addition to the realms of sense and motivation, a realm of culture can also be found in the anthropology of travel. It has a markedly different vocabulary, narratives, and foci that are disharmonious with the Western-middle-class centric paradigm of tourism studies. As such, this strand of scholarship has been not as oft-quoted as aforementioned conceptual constructs in the study of mass tourism. Nevertheless, it is critical to our understanding of Red Tourism for the work sutures culture, history, class, identity, and politics with travel. Clifford's notion of routes/roots is a prime example.

For Clifford (1992), cultures are sites of both dwelling (roots) and traveling (routes), all of which dispel the travel myth that tourists travel in unconstrained ways. Clifford contended that the touristic circuits and routes are determined by cultural and historical roots. In other words, different travelers (e.g., privileged bourgeois travelers, oppressed immigrant and migrant laborers, pilgrims, etc.) produce and follow different well-trodden routes. Clifford's anthropological view of travel calls for taking tourist discourse broadly and seriously on the one hand and reaffirms that tourism is a site of ideology with many layers on the other hand.

The first layer is culture. Clifford suggested that in some cases travel defines us culturally even more than dwelling does. At the very beginning of his influential essay *Traveling cultures*, Clifford (1992) quotes C.L.R. James' *Beyond a boundary* (1984), saying what really matters is "not where you are or what you have, but where you come from, where you are going and the rate at which you are getting there" (p. 96). Referring to the traveler as the "intercultural figure," Clifford regarded travel as a series of intercultural encounters inasmuch as it means to leave one group's core to one's periphery (another's core). This is why Appadurai (1996) referred to the global flow of travelers as "cultural traffic" (p. 28). Clifford went further, arguing that intercultural interpretation in such touristic encounters is necessarily politically charged. But travel is also productive, producing "cultural expressions" such as travelogues, comportments, books, and so forth (Clifford, 1992, p. 108).

Another layer is power. It is deeply interconnected with what Clifford (1992, p. 110) calls "historical taintedness" manifested in "gendered, racial bodies, class privilege, specific means of conveyance, beaten paths, agents, frontiers, documents, and the like." In Paul Gilroy's (1993) account of black Atlantic diaspora, it is political, racialized, and class power that constructed different types of traveling such as coerced travel by slaves, recreational travel by Whites, and the third type by the Pullman porter. In tourism studies, the dominant narrative of hegemonic tourist discourse led Bruner (1991) to argue that "[i]t (tourism) is as much a structure of power as it is a structure of meaning" (p. 240). Later, in examining historical tourism in Ghana, Bruner (1996) noted a considerable controversy over the representation of slavery in the castles for Dutch, British, and African American tourists. This is because each group of the tourists has a specific agenda predetermined by their interpretations of particular dimensions of history. To minister varied political interests, the local tour guide developed a strategy to take different groups of the foreign tourists to different places of the castles with different representations, thusly, different symbolic power. Sometimes such power can be brutal. For example, the powerful countries' fantasies about the Other have created loops of sex tours in Asian destinations (Appadurai, 1996). Sometimes that power is about class. In assessing different kinds of romantic tourism in England,

Walter (1982) has noted that most tourists are concentrated within a very limited area, whereas the privileged tourists consume the scarce places. Interestingly, Walter (1982) suggested that the tourists' choices of romantic places were largely due to persuasion of middle-class professional opinion-formers who were part of "an effective propaganda machine" (p. 300). This is to say, the popular routes of romantic tourism were partly a result of popular propaganda.

The third layer is education. It is in line with one of the characteristics of *xuanchuan*. In the *xuanchuan* model, to propagandize, to a great extent, is to educate. Historians have valorized the educative role of travel in the West, noting that travelers acquired certain knowledge by traveling certain routes. Meade (1914) described the continental tour in the 18th-century Europe as "an indispensable form of education for young men in the higher ranks of society" (p. 3). Meade suggested that although the tourist routes could be varied depending on personal interests and other factors, they did follow a pattern, generally conforming to the taste of Englishmen. To put it another way, the English aristocracy held hegemonic power in the discourse of the Grand Tour in terms of prescribed routes. In addition to foreign tours, domestic travels by British landed classes were also believed to be educational (Moir, 1964/2013). Like propaganda, the effects of travel as educational agency were incalculable; but undeniably, travel had become an important part of informal education in Britain by the end of the Industrial Revolution (Dent, 1975).

Overlapped with culture, power, and education, politics is another important layer of the route metaphor. Contemporary research in tourism, heritage tourism in particular, has demonstrated that the educational function of tourism can be highly political in a sense that the former educational agency now transforms into the propaganda machinery. This is why Hewison (1987) was strongly critical of what he terms the "heritage industry," denoting massively manufactured and widely distributed bogus history through heritage tourism. Hewison assumed some kind of propaganda-like anti-democratic feature with heritage tourism. Calling Hewison's argument "condescending," Urry (2002) disagrees because he sees heritage tourism as something like *xuanchuan*, propagandistic yet not necessarily mendacious. In examining the Wigan Pier, a heritage center in the UK, Urry (2002) argued that the site conveys "a non-elite popular culture," contended that the touristic attraction is "scholarly and educational," and concluded that compared to other media representations of history such as biographies, historical novels, and dramas, heritage tourism is no more misleading (p. 102). In a recent study of a dark tourism route in Northern Ireland, Skinner (2016) demonstrated that the walking tour is used to propagandize Republican ideology. Rather than manufacturing historical representations, or "staged authenticity," manipulating tourists' movement through the guided walking route became an effective means for such propaganda. I will elaborate the political dimension of tourism further in the next chapter as a history of political tourism in China unfolds.

Besides the theories of the organized tourist gaze, staged authenticity, and determined routes, other ideas in critical studies of tourism also point out that the site of tourism is ideologically powerful. Revealing the role of popular myth and fantasy in the sociocultural construction of tourist sight, for example, Rojek (1997) treated tourism as an "index" of representational cultures. According to this theory, a tourist sightseeing Notre Dame de Paris would also search for, in his mind, how it is "indexed" in popular culture, say, Victor Hugo's novel *The hunchback of Notre-Dame* or the film, to (re)construct the sight. Rojek stressed what he refers to as the "inner journey" of the tourist. An equivalent journey to the physical one in the inner world, inner journey designates a sense making process of the tourist through mobilizing popular cultural myths, or popular propaganda. For Rojek, the tourist experience involves two types of journeys to both physical space and abstract space. Together, they construct the touristic sight.

Mass communication: an analogy with journalism

Tourism is an "ism," a site of ideology. Theories of the organized tourist gaze, staged authenticity, and determined routes indicate that. Tourism is an ism roughly analogous to journalism. The two are similar in many respects including their persuasive/communicative nature, educational/informational role, and authenticity/objectivity concern by the audience. The analogy leads me to argue that tourism is a media system somewhat like the news media. However, there is a seemingly insurmountable gap in this comparison: journalism is a political institution (Cook, 2006; Domke, 2004; Nerone, 2015), but tourism seems not to be. The issue has to be addressed before I put tourism and journalism side by side for comparison.

The overriding problem about institution and structure of power can be surmised as this: journalism is a profession but tourism is not; or alternatively, reduced to a simple question: journalists produce journalism but do tourists produce tourism? It depends. From the perspective of industry, there is no such group of professionals that can be identified as *the* producer of tourism. There are many instead, including government officials, professionals from travel/tourism agencies, tour operators, parks, museums, and the like. They are equally important to the development of the tourism industry. What's more, there is no unified professionalism and norms for the professionals to embrace in tourism except for those for conducting regular business. All of these seemingly distance tourism as nothing like journalism, which is highly institutionalized with generally accepted norms and professionalism.

But the tourist is the producer of tourism from a cultural studies perspective. Distinct from other forms of popular culture such as television, film, popular music, and fiction, the core of tourism is spatial practice. The meaning of tourism is only delivered through tourists' spatial practices such as touring and gazing. The role of the tourist resembles that of what de Certeau (1984) metaphorically refers to as "the walker," who "transforms each spatial signifier into something else" (p. 98). Tourism is produced by the spatial practice of the tourist, not some kind of professional. Without tourists, tourism is meaningless and non-existent as there are only natural scenes and historical sites. The real producer of tourism is the tourist. This may help explain why tourism researchers have such a pervasive and persistent fascination with the tourist rather than the "tour" part or "ism" part of tourism.

The issue can be addressed differently, or interculturally. While tourism as a site of ideology does not have its formal institutions in the West, it is not necessarily the case in other parts of the world. In China, for example, tourism has a hierarchical system that parallels the nation's media system. The China National Tourism Administration (CNTA) is the organization in charge of tourism affairs at the central government level. Like the State Administration of Press, Publication, Radio, Film and Television (SAPPRFT), CNTA is directly affiliated to the State Council and also has its bureaus at the provincial level. The Office of National Red Tourism Coordination Group (ONRTCG), usually referred to as the "Red Office," is an internal department of CNTA in charge of Red Tourism. Interestingly, the head of the Red Office is not from CNTA, but the deputy director from National Development and Reform Commission (NDRC), a more powerful government agency responsible for planning and control over the Chinese economy on the macro level. The head of CNTA is the deputy director of the Red Office. Nevertheless, the organizational structure of the Red Office is much more complicated than that. Conference highlights of the Red Office show that in addition to NDRC and CNTA, the leadership of the Red Office comes from many other government departments and agencies consisting of the Publicity Department of the CPC (previously the Propaganda Department), the Ministry of Finance, Ministry of Education, the Ministry of Civil Affairs, the Ministry of Housing and Urban-Rural Development, the Ministry of Transport, China Railway, the Ministry of Culture, Civil Aviation Administration of China, and the State Administration of Cultural Heritage.

It now seems to become clear that given government control and institutionalization, tourism in China can be seen as a system similar to the country's news media system, but not elsewhere in the world. Tourism in Western democratic countries, for example, seems to be a self-organized, independent communication system insofar as there are no formal institutions and mechanisms to govern the tourists' behavior and tourists do not form any institution. However plausible, this is not what I mean. I do not want to make a special case of China's tourism against tourism in the rest of the world. Instead, I argue that tourism can be better understood, by and large, as a media system due to its enormous communicative power even if not institutional power. And this communicative power of tourism is comparable to that of journalism. Let me explain.

First, lacking formal institutional power does not mean lacking communicative power. Take the public sphere for instance. Theoretically constructed as a connecting space between the private and the state for the public to participate in political debates, the public sphere is believed to produce enormous communicative power for democracy even without being institutionalized. Put simply, the public sphere is self-organized. In light of the idea of the *xuanchuan*-public sphere as noted earlier, a similar discursive space or sphere produced in tourism by tourists exists. Tourists do not have their own institution, but a "sphere" instead. Translating this logic into de Certeau's (1984) language, tourists do not have "strategies" but they develop "tactics." Nevertheless, considering the bred-in-the-bone "frivolous" image (or imaginary in a Lacanian sense) of the tourist, hardly can anyone possibly argue that tourists in that sphere will act like the Habermasian public who characteristically engage in serious talk about a range of societal problems. Nevertheless, the communicative power of the "tourist sphere" cannot be underestimated either. The vast flows of tourists, ideas, cultural goods, and services along with the markers of class and lifestyles make tourism comparable to journalism in terms of massive flow of information.

And that information is not just any kind; it involves the kind that matters to public intelligence. At the heart of tourism or journalism is classic liberalism with a particular form of practice, whether free movement or free expression. What is extremely critical to these practices is what Nerone (2015) refers to as public intelligence, designating a combination of a special kind of information and the capacity for transforming individuals/people/self into the public. Public intelligence is so crucial to the liberal thinking of modern mass communication that it can thread through John Dewey's idea of community and Habermasian idea of the public sphere. While this pivotal role of journalism can be easily comprehended, linking tourism to public intelligence requires much deliberation, perhaps, much more imagination. This is, in part, why I started this chapter with a lengthy discussion on popular culture. In particular I have argued that popular culture is the popular site for propaganda. The great debate surrounding the culture industry also indicates that popular cultural industries like tourism matter a lot to public life. Still, it would be idle to compare tourism with journalism as much as to compare tourists with journalists. Tantalizing as this may sound, at least on one precise point such a comparison is very revealing.

That point is about the Real. It is "authenticity" in tourism and "objectivity" in journalism. One thing that tourism distinguishes from other popular cultural texts and practices is something one may call "touristic authenticity," a parallel description to journalistic objectivity in journalism. Touristic authenticity has two dimensions: the tourist experience (perception) and the site authenticity (objectivity). Wang (1999) argued that independent of the site objectivity, the perceived authenticity by the tourist is provoked by a "more authentic and more freely self-expressed" feeling activated in touristic encounters in the absence of constraints of everyday life (pp. 351–352). In other words, tourists see sights more authentic and real compared to their second-hand experiences. This is understandable considering the doctrine of "Seeing Is Believing." Similarly, news audiences also demand a certain degree of authenticity. In journalism studies, scholars usually refer to it as objectivity. Like authenticity to Europe-centric tourism studies, objectivity has been central to the development of the Western culture of journalism (Nerone, 2015; Schudson, 2001). Just as authenticity is the touristic experience in tourism, objectivity in journalism is merely the journalistic account of facts. Theoretically speaking, touristic authenticity is largely *staged* according to certain tactics (e.g., spatial manipulation, simulation/simulacrum), whereas objectivity in journalism is believed to be crafted by journalistic practices (e.g., news balancing, sourcing) or "rituals" in Tuchman's (1972) description. To put it simply, both authenticity and objectivity are not a matter of the Real, but a matter of reconstruction of reality within a society.

They point to a shared realm of mediated representation between tourism and journalism. They, then, become even more comparable in light of theories of representation. Recall that the theories (here in a broader sense as organized ways of thinking) in tourism propose that touristic representation is staged by spatial manipulation, perceived by organized ways of seeing, and even the touristic route to the representation highly determined by cultural and historical roots. Uncannily familiar, these theories do speak quite a bit for journalism too. Both staged authenticity and news framing, for example, describe the effective mechanisms for manipulating a given communicating text. News framing is about foregrounding some aspects of reality to communicate while omitting the rest (Entman, 1993; 2004; Gamson & Modigliani, 1989). This is just another way of saying "staging." In other words, framed reality in journalism can be explained by staged authenticity in tourism, and vice versa. Entman (1993) further notes that:

> [T]he frame determines whether most people notice and how they understand and remember a problem, as well as how they evaluate and choose to act upon it. The notion of framing thus implies that the frame has a common effect on large portions of the receiving audience, though it is not likely to have a universal effect on all.
>
> (p. 54)

To paraphrase the quoted paragraph: the news readers' gaze is organized by news frames. Hegemony and the indexing theory in media studies suggest that the making of news frames is not random but follows certain patterns consistent with the views of powerful elites (Gitlin, 1980; Hallin, 1987; Bennett, 1990; 2011). Translating this point into the language of tourism, news frames do follow highly determined routes whereby social and political power is distributed. In fact, many of the principal concepts of tourism studies, when translated, more or less correspond to classic concepts available on the reading list of the sociology of the press.

In spite of the similarities in the manipulation of representations, however, there is a glorious hiatus between tourism and journalism that cannot be touched on lightly. Notably, tourism does not have some fundamental functions that are expected from journalism, namely, "watchdog" and "gatekeeper." It reconfirms the fact that tourists are not journalists and have no institutional power. But beyond that, tourism is a powerful communication system similar to journalism at the macro level as both operate as a system for social integration. In studying journalism history, Nerone (2015) described the mechanism of that system as communication "by sifting through information in a way that can be presented publicly as both worthy of guiding policy and as the sort of thinking that really represents the way the public thinks" (p. 1). Curran (1978) has argued that what the system does is to propagate collective values and to legitimate the dominant ideology. In tourism studies, Matthews and Richter (1991) refer to this integrating function of tourism as political socialization. In light of all these, tourism can be seen as a propaganda system. Drawn from empirical data, my later narratives and analysis of the case of Red Tourism will shed additional light on this.

Conclusion

Treating tourism as a cultural practice, scholars have complicated the common notion of tourism as a business. Rojek and Urry (1997) pointed out a paradigm of cultural analysis while rejecting "a specifically social science of tourism" (p. 5). The treatment of tourism as a cultural practice, while bringing cultural thinking into the study of tourism, has an inherent deficiency that might prove to be its Achilles' heel: it renders the already vague and inclusive term "tourism" even much vaguer and all-embracing in light of being "cultural." A useful tactical move might be to re-contextualize the "cultural." This is what Rojek and Urry attempted. Nevertheless, they opted for the realm of human senses, which appears to be rather a devolution to Urry's earlier idea of the "tourist gaze" than a real "cultural" revolution. To tackle the problem, I have looked further back, revisiting the idea of the culture industry advanced by the scholars of the Frankfurt School from which the mass tourism industry stemmed.

Although never using the exact word "propaganda" in their narratives, Horkheimer and Adorno (2002; see also Adorno, 1991) do articulate the propagandistic nature of the culture industry by suggesting that it is deceptive and homogenous. They note that the public mentality is a part of the production of the culture industry. For Horkheimer and Adorno, the culture industry is an ideological filter through which once diverse contents and discourses were harmonized (they illustrated this point in the example of jazzing up Mozart) as dissent was blocked and removed, resulting in ideological hegemony. In the film industry, they referred to this filtering mechanism as "the guideline," a phenomenally popular term in the lexicon of propaganda. According to Horkheimer and Adorno, an important consequence of the culture industry is the dismantling of the public. They suggested that this process of propagandizing is somewhat like education, noting, "Donald Duck in the cartoons and the unfortunate victim in real life receive their beatings so that the spectators can accustom themselves to theirs" (Horkheimer & Adorno, 2002, p. 110). Nevertheless, scholars from the field of cultural studies, particularly those from the Birmingham School assumed a more intelligent and critical-thinking capable public than the one in Horkheimer and Adorno's analysis (e.g., Hall et al., 1980; Hall & Jefferson, 1975).

The vigorous debates on the culture industry have been inevitably passed onto the tourism industry with a similar point of contention: the gullibility of the tourist. This is eerily reminiscent of the strong-*versus*-weak/non-effect debate in early media effects studies and propaganda studies. But such debates have less to do with my view of tourism as a media/propaganda system because whether tourists are smart or gullible is independent of the presence of such a system.

There is an enormous gap between tourism studies and propaganda studies. For one, analysis of ideology and politics has been notably absent in tourism studies;

for another, propaganda researchers' marked preferences for the news media and communication systems leave almost no space for any serious inquiry into tourism. Also, systematically lacking in social/critical studies of tourism is the role of state. This is in part because scholarship on tourism is Anglo-Western centric and the tourism industry in the Western democratic countries is commonly seen as a laissez-faire business, private and individualized (Matthews & Richter, 1991). As such, discussion of politics in tourism development is normally limited to the level of local authorities, without considering much about state power, let alone state propaganda. To fill the gap theoretically, I proposed the TPF. It is a mass communication-oriented theoretical construct remodeled from the Congregatio de Propaganda fide, the archetypical system of propaganda in the West with the substitutes of missionaries for tourists, of religion for ideology. At the core of both systems is a communicatively powerful form of spatial practice, characterized by human mobility and social communication. As a critical mode of tourism inquiry, though preliminary and lacking in sophistication, the TPF innovatively connects three different modes of thinking, namely, the public sphere, the culture industry, and the practice of propaganda, integrating the three discursive spaces into a triumvirate-sphere of popular culture in which mass tourism is viewed as a mass propaganda system.

To end this chapter, an anecdote from Mao Zedong would be illuminating. In 1944, in conversation with US Foreign Service officer John S. Service on the topic of propagandizing American democracy in China, Mao said, "every American soldier in China should be a walking and talking advertisement for democracy" (Davies, 2012, p. 216; Esherick, 1974, pp. 302–303). This is the first part of the story, which has been widely circulated as evidence of Mao's early positive attitude toward Western democracy. More revealing, perhaps, is the rarely told second part. According to historical records published in an internal publication of the party, the story then goes like this: sensing the flavor of propaganda, Service rejected Mao's proposal, stating that the US Army cannot work as a propaganda force. Then Mao explained what he meant was not to let the American soldiers do any propaganda work, but simply to be there presenting themselves (Party History Research Center, 1983). For Mao, the presence of the American travelers themselves is propagandistic enough. The point is that the traveler can be a powerful agent for propaganda, and presumably, tourism can be a propaganda agency. This particular way of thinking about tourism, the treatment of tourism as a media system, is one of the focal points of this book, and the subsequent chapters are meant to illustrate this point. Moving from theories of propaganda and tourism, I will present a brief social history of political tourism in China in the next chapter within the TPF by stressing those overlooked yet critical elements in the study of tourism, namely, the state, ideology, and politics.

Note

1   It involves five stages: preservation, framing and elevation, enshrinement, mechanical reproduction, and social reproduction. For a detailed description of multistage sight sacralization, see MacCannell (1999), pp.44–45.

References

Adorno, T. (1975). Culture industry reconsidered. *New German Critique, 6*, 12–19.

Adorno, T. (1991). *The culture industry: Selected essays on mass culture*. Routledge.

Apostolopoulos, Y., Leivadi, S., & Yiannakis, A. (Eds.). (1996). *The sociology of tourism*. Routledge.

Appadurai, A. (1996). *Modernity at large: Cultural dimensions of globalization*. University of Minnesota Press.

Aramberri, J. (2010). *Modern mass tourism*. Emerald.

Auslander, H. B. (1981). "If Ya Wanna End War and Stuff, You Gotta Sing Loud"—A survey of Vietnam-related protest music. *Journal of American Culture, 4*(2), 108–113.

Benjamin, W. (2008). *The work of art in the age of its technological reproducibility, and other writings on media*. The Belknap Press of Harvard University Press.

Bennett, W. L. (1990). Toward a theory of press-state relations in the United States. *Journal of Communication, 40*(2), 103–125.

Bennett, W. L. (2011). *News: The politics of illusion*. Longman.

Boorstin, D. (1992). *The image: A guide to pseudo-events in America*. Vintage.

Bruner, E. M. (1991). Transformation of self in tourism. *Annals of Tourism Research, 18*, 238–250.

Bruner, E. M. (1996). Tourism in Ghana: The representation of slavery and the return of the black diaspora. *American Anthropologist, 98*(2), 290–304.

Chan, Y. W. (2006). Coming of age of the Chinese tourists: The emergence of non-Western tourism and host–guest interactions in Vietnam's border tourism. *Tourist Studies, 6*(3), 187–213.

Chinchilla, R. (2020, July 22). *"Unacceptable", "unsafe": Philly officials back removal of Columbus statue*. NBC. https://www.nbcphiladelphia.com/news/local/unacceptable -unsafe-philly-officials-back-removal-of-columbus-statue/2476428/

Clifford, J. (1992). Travelling cultures. In L. Grossberg, C. Nelson & P. Treichler (Eds.), *Cultural studies* (pp. 96–116). Routledge.

Cloke, P., & Perkins, H. C. (1998). "Cracking the canyon with the awesome foursome": Representations of adventure tourism in New Zealand. *Environment and Planning D: Society and Space, 16*(2), 185–218.

Cohen, E. (1972). Toward a sociology of international tourism. *Social Research, 39*(1), 164–182.

Cohen, E. (1988). Traditions in the qualitative sociology of tourism. *Annals of Tourism Research, 15*(1), 29–46.

Cohen, E. (1995). Contemporary tourism – trends and challenges: Sustainable authenticity or contrived post-modernity? In R. Butler & D. Pearce (Eds.), *Change in tourism: Peoples, places, processes* (pp. 12–29). Routledge.

Cook, T. E. (2006). The news media as a political institution: Looking backward and looking forward. *Political Communication, 23*(2), 159–171.

Curran, J. (1978). The press as an agency of social control: An historical perspective. In G. Boyce, J. Curran & P. Wingate (Eds.), *Newspaper history from the seventeenth century to the present day* (pp. 51–75). Constable.

Dash, G., & Cater, C. (2015). Gazing awry: Reconsidering the Tourist Gaze and natural tourism through a Lacanian–Marxist theoretical framework. *Tourist Studies, 15*(3), 267–282.

Davies, J. P. (2012). *China hand: An autobiography*. University of Pennsylvania Press.

de Certeau, M. (1984). *The practice of everyday life*. University of California Press.

DeFleur, M. H. (1998). James Bryce's 19th century theory of public opinion in the contemporary age of new communication technologies. *Mass Communication & Society*, *1*(1–2), 63–84.

Dent, K. (1975). Travel as education: The English landed classes in the eighteenth century. *Educational Studies*, *1*(3), 171–180.

Domke, D. S. (2004). *God willing: Political fundamentalism in the White House, the "war on terror" and the echoing press*. Pluto Press.

Entman, R. M. (1993). Framing: Toward a clarification of a fractured paradigm. *Journal of Communication*, *43*(4), 51–58.

Entman, R. M. (2004). *Projections of power: Framing news, public opinion, and US foreign policy*. University of Chicago Press.

Esherick, J. W. (Ed.). (1974). *Lost chance in China: The World War II despatches of John S. Service*. Random House.

Feighery, W. G. (2012). Tourism and self-orientalism in Oman: A critical discourse analysis. *Critical Discourse Studies*, *9*(3), 269–284.

Fico, F., & Cote, W. (1999). Fairness and balance in the structural characteristics of newspaper stories on the 1996 presidential election. *Journalism & Mass Communication Quarterly*, *76*(1), 124–137.

Fiske, J. (1987). *Television culture*. Routledge.

Foucault, M. (1975). *The birth of the clinic: An archaeology of medical perception*. Vintage.

Foucault, M. (1977). *Discipline and punish: The birth of the prison*. Vintage.

Foucault, M. (1978). *The history of sexuality, volume 1: An introduction*. Pantheon.

Foucault, M. (1988). *Madness and civilization*. Vintage.

Franklin, A. (2001). *The tourist gaze* and beyond: An interview with John Urry. *Tourist Studies*, *1*(2), 115–131.

Fryer, P. (1984). *Staying power*. Pluto.

Gamson, W., & Modigliani, A. (1989). Media discourse and public opinion on nuclear power: A constructionist approach. *The American Journal of Sociology*, *95*(1), 1–37.

Gilroy, P. (1993). *The black Atlantic: Modernity and double consciousness*. Verso.

Gitlin, T. (1980). *The whole world is watching: Mass media in the making and unmaking of the new left*. University of California Press.

Goffman, E. (1956). *The presentation of self in everyday life*. University of Edinburgh Social Sciences Research Centre.

Graburn, N. H. H. (1977). Tourism: The sacred journey. In V. Smith (Ed.), *Hosts and guests: The anthropology of tourism* (pp. 17–31). University of Pennsylvania Press.

Greenwood, D. J. (1972). Tourism as an agent of change: A Spanish Basque case. *Ethnology*, *11*(1), 80–91.

Habermas, J. (1989). *The structural transformation of the public sphere: An inquiry into a category of bourgeois society*. MIT Press.

Hall, S., & Jefferson, T. (Eds.). (1975). *Resistance through rituals*. Hutchinson.

Hall, S., Hobson, D., Lowe, A., & Willis, P. (1980). *Culture, media and language*. Heinemann.

Hallin, D. (1987). Hegemony: The American news media from Vietnam to El Salvador, study of ideological change and its limits. In D. Paletz (Ed.), *Political communication research* (pp. 3–25). Ablex.

Herman, E. S., & Chomsky, N. (1988/2008). *Manufacturing consent: The political economy of the mass media*. Pantheon.

Hewison, R. (1987). *The heritage industry: Britain in a climate of decline*. Methuen.

Horkheimer, M., & Adorno, T. (2002). *Dialectic of enlightenment*. Stanford University Press.

Huang, P. C. (1993). "Public sphere"/"civil society" in China?: The third realm between state and society. *Modern China, 19*(2), 216–240.

James, C. L. R. (1984). *Beyond a boundary*. Pantheon.

James, D. (1989). The Vietnam War and American music. *Social Text, 23*, 122–143.

Katz, E. (2006). Rediscovering Gabriel Tarde. *Political Communication, 23*(3), 263–270.

Klein, C. (2003). *Cold War orientalism: Asia in the middlebrow imagination, 1945–1961*. University of California Press.

Klein, M. (1990). Historical memory, film, and the Vietnam era. In L. Dittmar & G. Michaud (Eds.), *From Hanoi to Hollywood: The Vietnam War in American film* (pp. 19–40). Rutgers University Press.

Kracauer, S. (1995). *The mass ornament*. Harvard University Press.

Lazarsfeld, P. F., Berelson, B., & Gaudet, H. (1944). *The people's choice*. Duell, Sloan, & Pearce.

Lefebvre, H. (1991). *The production of space*. Blackwell.

Liang, Z. P. (2003). "Minjian," "minjian shehui" he CIVIL SOCIETY --CIVIL SOCIETY gainian zai jiantao [Reconsidering the concepts of "civil" and "civil society" in the Chinese context]. *Yunnan daxue xuebao, 1*, 56–68.

MacCannell, D. (1973). Staged authenticity: Arrangements of social space in tourist settings. *American Journal of Sociology, 79*(3), 589–603.

MacCannell, D. (1999). *The tourist: A new theory of the leisure class*. University of California Press.

MacCannell, D. (2001). Tourist agency. *Tourist Studies, 1*(1), 23–37.

Maoz, D. (2005). The mutual gaze. *Annals of Tourism Research, 33*(1), 221–239.

Marcuse, H. (2002). *One dimensional man*. Routledge Classics.

Matthews, H. G., & Richter, L. K. (1991). Political science and tourism. *Annals of Tourism Research, 18*(1), 120–135.

Meade, W. E. (1914). *The Grand Tour in the eighteenth century*. Houghton Mifflin.

Moir, E. (1964/2013). *The discovery of Britain: The English tourists 1540–1840*. Routledge.

Nerone, J. (2015). *The media and public life: A history*. Polity.

Party History Research Center (Ed.). (1983). Mei yige zai zhongguo de meiguobing dou yingdang chengwei minzhu de huoguanggao [Every American soldier in China should become a living advertisement for democracy]. *Dangshi tongxun*, 20–21.

Pearce, P. L. (1982). *The social psychology of tourist behavior*. Pergamon.

Potuoglu-Cook, O. (2006). Beyond the glitter: Belly dance and Neoliberal gentrification in Istanbul. *Cultural Anthropology, 21*(4), 633–660.

Rojek, C. (1997). Indexing, dragging and the social construction of tourist sights. In C. Rojek & J. Urry (Eds.), *Touring cultures: Transformations of travel and theory* (pp. 52–74). Routledge.

Rojek, C. (2000). Mass tourism or the re-enchantment of the world? Issues and contradictions in the study of travel. In M. Gottdiener (Ed.), *New forms of consumption: Consumers, culture, and commodification* (pp. 51–70). Rowman & Littlefield.

Rojek, C., & Urry, J. (1997). Transformations of travel and theory. In C. Rojek & J. Urry (Eds.), *Touring cultures: Transformations of travel and theory* (pp. 1–19). Routledge.

Said, E. (1977). *Orientalism*. Penguin.

Said, E. (1985). Orientalism reconsidered. *Cultural Critique, 1*, 89–107.

Schudson, M. (2001). The objectivity norm in American journalism. *Journalism, 2*(2), 149–170.

Shields, R. (1990). *Places on the margin*. Routledge.

Skinner, J. (2016). Walking the falls: Dark tourism and the significance of movement on the political tour of West Belfast. *Tourist Studies, 16*(1) 23–39.

Storey, J. (2009). *Cultural theory and popular culture* (5th ed.). Pearson Education.

Sun, Lung-kee. (2000). *Zhongguo wenhua de shenceng jiegou* [The deep structure of Chinese culture]. Guangxi shifan daxue chubanshe.

Tarde, G. (2010). *Gabriel Tarde on communication and social influence*: Selected papers. University of Chicago Press.

These Confederate statues were removed. But where did they go. (2020, September 21). NBC. Retrieved October 16, 2021 from https://www.nbcnews.com/news/us-news/these-confederate-statues-were-removed-where-did-they-go-n1240268

Tuchman, G. (1972). Objectivity as strategic ritual: An examination of newsmen's notions of objectivity. *American Journal of Sociology, 77*(4), 660–679.

Urry, J. (1992). The tourist gaze "revisited." *American Behavioral Scientist, 36*(2), 172–186.

Urry, J. (2002). *The tourist gaze*. Sage.

Walter, J. (1982). Social limits to tourism. *Leisure Studies, 1*(3), 295–304.

Wang, N. (1999). Rethinking authenticity in tourism experience. *Annals of Tourism Research, 26*(2), 349–370.

Williams, R. (1960). *Culture and society: 1780–1950*. Anchor.

Williams, R. (1980/1997). *Problems in materialism and culture*. Verso.

Williams, R. (1989). *Resources of hope: Culture, democracy, socialism*. Verso.

Williams, R. (2003). *Television*. Routledge Classics.

Xia, J. F., & Huang, Y. Q. (2008). "Gonggong lingyu" lilun yu zhongguo chuanmei yanjiu de jiantao: tanxun yizhong guojia—shehui guanxi shijiao xia de chuanmei yanjiu lujing [A critical study of public sphere theory and Chinese media research: Searching for a media studies method from a state-society perspective]. *Xinwen yu chuanbo yanjiu, 5*, 37–46.

Yan, C., & Santos, C. A. (2009). "China, Forever": Tourism discourse and self-orientalism. *Annals of Tourism Research, 36*(2), 295–315.

4 The cultural roots of Red Tourism

Culture has countless meanings. Williams proposed a triad framework for analysis of culture. For Williams (1998), culture is a dynamic process that takes place in three dimensions: the ideological dimension ("ideal"), the historical dimension ("documentary"), and the social dimension ("a particular way of life"). Therefore, for Williams, a comprehensive analysis of culture should be partly ideological, partly historical, and partly social. I apply this analytical framework for my historical account of Red Tourism of which the culture of *xuanchuan* is part.

Instead of treating Red Tourism as a type of political tourism, I see it as what I call "*mass* political tourism." It distinguishes the existing idea of "political tourism" with an added "mass." Mass herein has two specific meanings: first, it denotes the massive space that tourists travel; second, it designates the main body of the tourists as the mass. In other words, the majority of the tourists engaging in political tours in China are not partisans (e.g., leftists, conservatives, party members) but regular tourists like you and me and they travel all over the country.

My insistence on using "mass" reflects the general argument of this chapter: political tourism in China is not a conglomeration of many types of political tours. Rather, Chinese political tourism has been a well-established social institution throughout Chinese history. Political tourism in China has its own guiding philosophies, bureaucratic institutions, and communication pathways. Seen as "dark tourism" in Eastern Europe (Light, 2000), communist heritage tourism has evolved into an industry only in China. Also, the Chinese tourists' perception of Red Tourism is generally positive (Hu, 2009). Together, it suggests a different paradigm of political tourism where the mass matters.

This chapter is a historical study of Red Tourism. Against a backdrop of a Chinese philosophy of tourism, it explores how Red Tourism originated, evolved, and transformed from the separate ideas of "Red" and "Tourism" into a Red Tourism industry with an overwhelming emphasis on political thinking along with the emergence of institutional power. The chapter has three purposes. The first is to bring the interplay between tourism and propaganda, as I have discussed in previous chapters, from theory to practice; the second is to unravel the political orientation of Chinese tourism and how mass political tourism has been institutionalized in the case of Red Tourism; the third purpose is to reveal what "ism" is behind the powerful rise of Red Tourism.

DOI: 10.4324/9781003231783-4

Three sources inform this study. First and foremost, I drew from my own fieldwork in multiple locations in China between 2011 and 2021, an extensive data collection containing archives, collected documents, notes, photographs, and interviews. A total of 16 in-depth interviews were conducted. For this chapter, I drew particularly from five respondents who were senior officials with government agencies and bureaus responsible for Red Tourism. All the interviewees' names have been anonymized throughout. Second, I analyzed popular discourse on political tourism in China. In examining the government discourse of Red Tourism, I delved into reports, editorials, and commentaries that I culled from the *People's Daily*. The newspaper was selected because the *People's Daily* represents the viewpoints of the Central Government (Wu, 1994). Third, supplemental to primary sources, Chinese scholarship in tourism studies and what some Chinese scholars call "Red Tourism studies," which has been largely unknown in the West, were also examined to shed some light on mass political tourism.

My narrative is organized chronologically. It starts from what I refer to as a "prehistory" of tourism in imperial China. It traces back Chinese philosophical ideas about "tourism," a social practice that is different from what we mean by "tourism" today but has significantly contributed to the popular imagination of tourism conjured by modern Chinese intellectuals. Then it follows a brief history of the modern Chinese tourism industry in Republican China. It is considered "brief" not so much as I wrote briefly as the emerging industry only survived less than three decades. The majority of this chapter concentrates on the PRC period since the establishment of the party-state in 1949. This is a period where a recent history of Red Tourism is located. Based on ideological transformations of tourism, I divided the PRC period into four sub-periods, namely, the New China period (1949–1965), the Cultural Revolution period (1966–1976), the early reform period (1977–1989), and the post-1989 period (1990–present) with an emphasis on the last one.

An overlapped pre-history of tourism with *xuanchuan*

The history of Chinese tourism, though written by different authors, reads partly like a history of Chinese philosophy, partly like a history of Chinese literati, partly like a history of Chinese emperorship, but nothing like a typical tourist-centered modern tourism study. Nevertheless, I would argue that it is the triad of the three principal roles, philosopher, literati, and emperor that defines the identity of Chinese tourism as something truly serious, powerfully communicative, and critically informative, all of which differ substantially from the modern notion of tourism as something personally pleasurable. It is something that renders early Chinese tourism essentially propagandistic. It is not an exaggeration that an ancient history of Chinese tourism is roughly equivalent to that of Chinese propaganda. Since tourism in imperial China mainly refers to travels made by social elites, which is different from modern tourism, I would call this part of history of Chinese tourism a pre-history.

The philosophical foundations

Let's start with the philosophers. The connection between tourism and *xuanchuan* manifests in the fact that the first generation of propagandists, commonly known as "Pre-Qin philosophers," were also considered the first group of tourism theorists. For example, Li and Zhao (2001) regarded Confucius and Zhuang Zi as the founders of Chinese tourism theory. Confucius believed the ultimate goal of tourism was about "self-cultivation, building family, ruling a state, and conquering the world" (*xiushen* 修身, *qijia* 齐家, *zhiguo* 治国, *ping tianxia* 平天下), which sounds exactly like *xuanchuan* (Wang & Zhang, 1998, p. 88). As a matter of fact, this is how Wang (1998) framed his historical narrative of tourism in ancient China, which is partly a history of emperorship and partly a history of conquering by means of intelligence.

Peng (2006) has generalized five philosophical ideas of tourism in Chinese history. The five strands of thought on tourism deal with five philosophical propositions on critical matters, specifically, morality, filial piety, the meaning of practice, absolute freedom, and the meaning of life. This is why a history of China's tourism, to a great extent, reads like a history of Chinese philosophy. Peng (2006) claimed in regards to tourism "Confucianism was about governing the state (*ru zhi guo* 儒治国), Buddhism about governing the soul (*fo zhi xin* 佛治心), and Taoism about governing the body (*dao zhi shen* 道治身)" (p. 14).

Two schools of thought are of paramount importance in later developments and perceptions of tourism. The first strand of thought from Confucius holds that tourism is a way to self-cultivation in that travelers can learn the cardinal virtues from the natural environment if they decipher the imaginative meaning in a figurative sense. It became what later was called the "figural virtue theory" (*bi-de shuo* 比德说). Many Chinese scholars hold that Confucius' thought on tourism was utilitarian since for Confucius, the goal of tourism was to help achieve political ideals (Fan et al., 2008). But Zhang (2008) argued that the Confucian view of tourism was an amalgam of the rational and the emotional.

In contrast, the second philosophical idea from Zhuang Zi 庄子 (369–298 B.C.E.) seems fairly emotional. It assumes that like meditation, tourism is meant to totally free self from reality in order to obtain Tao. The theory was named as "carefree wandering" (*xiaoyao you* 逍遥游). It has been said indisputably that Taoism opposes Confucianism in this regard. Nevertheless, I would argue that Confucius and Zhuang Zi actually shared the same idea that tourism was an optimal means of self-improvement with the only difference being that one preferred to interpret nature in relation to human nature and the other was inclined to let the feeling freely flow to "activate" the human nature. Both Confucius and Zhuang Zi's theories were ideology oriented, as they deal particularly with worldviews. And both can be seen as propaganda because they worked on propagating their worldviews.

Referring to the two philosophical ideas as "Confucian tourism theory" and "Taoist tourism theory," Wang and Zhang (1998) argued that however different, the two theories have influenced and guided Chinese tourism practice ever since.

In academia, Chinese scholars have consistently studied tourism from the lens of philosophy as "throughout history, what they have been interested in are the dialogues between the tourist and nature, historical relics, and inquiries into the sentiment relations between human beings and Heaven and Earth" (Yu, 2005, p. 2). In that regard, literati played a primary role.

The literati concentration

Tourism in ancient China, by all accounts, was literati-centered. Referring to *shi* 士 as the "intelligentsia group" (*zhishi qunti* 知识群体), Peng (2006) pointed out that the *shi* consisted of the majority of tourists in ancient time. There are many terms in ancient Chinese written records that were associated with tourism, for example, *yun you* 云游 (wander), *huan you* 宦游 (travel made by officials), *xun you* 巡游 (travel made by emperor), *xuan you* 玄游 (travel made by neo-Taoists), *xian you* 仙游 (travel for personal immortality), *shi you* 释游 (travel made by Buddhists), *shi lv* 士旅 (travel for securing an official position), etc. Each of these terms represents an ideology or a philosophy advocated by a particular group of Chinese literati. It is evident that the tourists in imperial China, though motivated by a variety of needs, whether for the pursuit of fame or for a sense of escapism, were largely depicted as the well-educated. Calling their work "a history of pathways combining those of acquisition and travel," Wang and Zhang (1998) suggested that tourism was an inseparable part of the Chinese literati life, preserving their identity as intelligentsia.

The early history of Chinese tourism is a necessary part of that of *xuanchuan*. In the Spring and Autumn Period, developments in transportation, trade, and the formation of early trade-centered cities along with the postal network led to the boom of *xuanchuan* practice and also the rise of tourism. Tourism as a social practice has intertwined with *xuanchuan* activities ever since then. *Kese* 客舍, a type of hostel provided by wealthy aristocrats to accommodate lobbyists (*shiren*, 士人), coexisted with the prototype of modern hostel for regular travelers. Wang (1998) noted that the most active tourists during this period were *shi*, who traveled to persuade those with power, mostly monarchs, to adopt their ideas. Chinese historian Qian Mu 钱穆 (1895–1990) (2001) characterized this period as the "lobbyist society" (*shoushi shehui* 游士社会) by the same token. Persuasion was the lobbyists' profession and their achievements could only be assessed by the effects of their *xuanchuan* activities. Because of that, Wang and Zhang (1998, p. 79) called tourism made by *shi* "utilitarian tourism." In sum, a Chinese history of *xuanchuan*/propaganda necessarily involves the power relations between the powerful and the learned, whereas tourists played a role. From here, propagandists in Chinese minds had been typically associated with intelligentsia, imagined as educated, intelligent, usually mastering literature, and itinerant. The social image of propagandist as dishonest and insincere political actors in China is rather contemporary.

To prepare themselves for a career in the persuasion business, *shi* first need to master their craft, which also required frequent travels. In the grand narrative of

Chinese tourism, Confucius alongside his preaching was credited as "improving the taste and aesthetics of tourism" to the degree that pupils not only learned to appreciate natural scenes during their journey but also adopted a habit to explore the philosophy of life beyond the beauty (Wang & Zhang, 1998, p. 78).

The literary legacy of Chinese tourism also points to a realm of Chinese literati. "Landscape literature" (*shanshui wenxue* 山水文学) is a unique and representative genre of ancient Chinese literature, literally "literature of mountains and waters." It developed in accordance with the development of tourism in China and became a "representation of space" of Chinese tourism, to borrow a term from Lefebvre. In the eyes of Chinese literati, landscape never stands alone but always signifies some kind of morality. Traditionally, Chinese people believe each natural phenomenon, a mountain or a river, articulates some aspects of human virtues that the tourist can learn from.

The emperorship association

Like propaganda, tourism in ancient China was also associated with emperorship and governance. Legends say, the Yellow Emperor (*Huangdi* 黄帝), the creator of the Chinese people, was a great traveler who visited numerous great mountains in his life. Because of the cognitive and cultural activities associated with his tours, the Yellow Emperor was deemed as the "seeder" of Chinese touristic culture by Chinese tourism historians (Wang & Zhang, 1998; Zhang, 1992).

Tourism intersected emperorship at a particular form of tourism, *Feng-Shan* 封禅. It is a combination of the worship ceremonies of the Heaven, which is *feng*, and of the Earth, *shan*, conducted by monarchs (Feng, 2007). While the procedures of the *Feng-Shan* Ceremony were different and vague in historical records, it has been agreed that the long, grand journey of the monarch to the ceremony site was just as propagandistic as the ritual ceremonies for legitimating monarchical power themselves, for it created and spread out an image of prosperity to a large proportion of the populace on that journey (Wang & Zhang, 1998). Emperor Qin Shi Huang 秦始皇 (259–210 B.C.E.), for example, propagandized his virtues and merits by carving stones on his journeys, functioning like some kind of premodern propaganda pamphlet or the iconography of propaganda to be more specific. This practice is eerily reminiscent of the Long March, which Mao famously called the "propaganda team" (*xuanchuan dui* 宣传队).

In imperial China, people perceived tourism as means of acquiring knowledge, of self-cultivation, of establishing worldviews for religious, political, ritual, literary, aesthetic, and educational purposes. But fun, the inerasable marker of modern tourism, was not on that long list however defined. In fact, stringing together all kinds of Chinese philosophical thinking on tourism points to a ubiquitous idea about a tourism that fundamentally opposed traveling purely for leisure. The Qin Empire declared that only if the hostel business (*nilv* 逆旅) was abolished would the development of agriculture be achieved. It reasoned that the business harbored a lot of "idle tourists" who could be engaged in farming otherwise (see Peng, 2006, p. 132). It seems that in the age of empires outside of the secular realm, only the social elites, whether literati or officials, were encouraged to travel. Perhaps, this is why tourism at the grassroots has gone mostly unmentioned in both historical records and modern retellings, or at the most, framed as celebrations of holidays along with other rituals and ceremonies that somehow involved travel.

The advent of modern tourism and its ideological articulation, 1911–1949

Modern Chinese tourism started as an exercise of patriotism in modern times. It was the embodiment of a political articulation of a mixed reality of revolutionary past and colonial present in a form that the Chinese people believed partly cultural and partly modern. This was also how Republican China (1911–1949) represented itself. Although influenced by the Western practice, the development of modern tourism in China was believed to be based on traditional Chinese culture with a theme of patriotism (Jia, 2002; Wang, 2005). This is to suggest that the keywords of early modern Chinese tourism are culture and patriotism. Together, they point to a realm of ideology, the inner core of mass political tourism.

The framing of this particular history of tourism is, too, politics-oriented. In *The modern history of China's tourism*, Wang Shuliang (2005) broke down the modern development of tourism in Republican China into three stages: the first stage was the formation and founding period, from 1912 to 1927; the second stage was the early development period, from 1927 to 1937; and the third stage was the setback and recovery period, from 1937 to 1949. Note that the four critical moments in this schema, namely, 1912, 1927, 1937, and 1949, represent the end of the Xinhai Revolution, the birth of the KMT regime, the outbreak of the Second Sino-Japanese War, and the establishment of the PRC. Except for the Anti-Japanese War, the other three moments connote three political revolutions, namely, the Revolution of 1911, the National Revolution, and the Communist Revolution. In that sense, Wang's periodization was rather revolution-based and political. Nevertheless, I am not suggesting that this is wrong; rather, I simply spell out the fact that the study of the modern history of tourism in China, to a great extent, connects to the development of the nation-state and national identity.

Wang and Zhang's (2003) account of the early Chinese tourism industry reads like a modern history of Chinese nationalism/patriotism. The narrative starts with depicting a group of Chinese students who studied abroad and later came back to China to engage in social reforms and it ends up with an account of a series of global travels made by Tao Xingzhi 陶行知 (1891–1946), a distinguished Chinese educationalist, in an effort to counter Japanese war propaganda. Overall, the work seems to suggest that the critical social changes of early modern China, to some extent, were brought about by those patriotic travelers/intellectuals. Unsurprisingly, like the ancient history of Chinese tourism, its modern parallel has still exclusively focused on social elites, intellectuals in particular, and how the substantial majority, the masses, traveled remains largely unknown. This is to say, tourism has been treated as a unique dimension of intellectuals throughout Chinese history, if not a form of narcissism.

The starting point of modern mass tourism in China has been narrated as a single patriotic act of a Chinese businessman. Chen Guangfu 陈光甫 (1881–1976), an Ivy League educated Chinese banker in Shanghai, created China Travel Service, the first Chinese-run travel agency in China in 1923 in response to those foreign travel agents who allegedly treated him with disdain, telling Chen that Chinese people were incapable of providing tourism service (Gross, 2011; Tong, 2009; Yao & Cao, 2002; Wang, 2005; Wang & Zhang, 2003). Upon its establishment, the tourism company aimed at spreading the nation's prestige (Peng, 2006; Wang & Zhang, 2003; Yi, 2009).

The business of the Travel Service was not good in the first few years. The reason was rather ideological. As noted earlier, Confucianists opposed tourism for pure leisure, claiming that travel must follow right principles (Fan et al., 2008). This anti-leisure ideology of tourism was so influential among the Chinese literati and intellectuals that it had become one of the tenets of Chinese tourism. The biggest challenge among what Gross (2011) referred to as "three impediments" to the early development of mass tourism in China was to persuade the urban middle class that there was nothing wrong about travel for fun and leisure was "appropriate" to their social status. In other words, the major task for the early Chinese tourism industry was not so much as to build up a primary clientele as to dismantle the hostile Confucian principle inherent in the Chinese culture of tourism. To do so, the travel agency used travelogues to sell the modern idea of tourism, branding it as a kind of representation of modernity. Acknowledging the propaganda power of tourism, Chen Guangfu explained to his employees that the seemingly unprofitable business would help his bank generate greater publicity, which he referred to as "invisible yields" (Wang, 2005, p.6). Nevertheless, patriotism never was off the scene of the newly developed Chinese tourism industry. In 1933, Chen's tourism company asked American journalist Edgar Snow to write five travel pamphlets in English about China as anti-Japanese propaganda and delivered 20,000 copies to the Chicago World's Fair through an American-Chinese association (Wang, 2005; Wang & Zhang, 2003). This particular way of thinking of tourism as a means of propaganda by ancient Chinese finally found its modern inheritor in a world of banknotes and stocks.

Central to patriotism was nationalism. According to a news report, tours of northeastern China had been increasingly gaining popularity in Shanghai right before the Anti-Japanese War, attracting lots of southeastern people thanks to its "historical richness" and "Chineseness" (Yousheng, 1936). In 1936, a few months before the start of the War, Yousheng, a Shanghai-based tourism company, having successfully organized two fully booked tours in a single year, continued to operate the 17-day tour in October. In response to overheated concern over the deteriorating military situation in the northeast, the company promised that they would have the military and police authorities to protect the tourists throughout the tour. The year of 1936 turned out to be a golden year for China Travel Service too as its revenue increased to five times the previous year's (Yi, 2009).

Nationalism was also exploited as a strategy to promote Huangshan, a scenic spot 250 miles southwest of Shanghai. Articles published in *China Traveler*, the first travel magazine in China, propagated the idea that touring Huangshang "would make the tourist strong in mind and spirt to prevail over foreigners" on the one hand and reaffirm their Chinese identity on the other (Gross, 2011, p.138). Nationalism was also packed in tourism related advertisements. A hotel advertisement written by Hu Shi 胡适 (1891–1962), a prominent writer and scholar, says, "New Asia Hotel's success has convinced us that our nation is not unable to live a clean and tidy life" (see Peng, 2006, p. 200).

The rediscovery of the propaganda power of tourism in the wake of WWII marked modern Chinese tourism differently from an industry point of view. The tourism industry had fairly developed when Republican China recovered from the War. Now tourism assumed three big tasks: (1) to educate citizens by fostering nationalism and patriotism through historical tourism, (2) to increase profits through inviting more foreign tourists, and (3) to propagandize the nation and its morality in order to improve foreign relations (Wang, 2005, p. 77). One specific task of post-war Chinese tourism was to preserve major Anti-Japanese War sites, including Lugouqiao 卢沟桥, Taierzhuang 台儿庄, and Changsha 长沙, "to commemorate the War of Resistance and to let the Chinese people be vigilant about the war, never forgetting those who sacrificed their lives in exchange for freedom" (Wang, 2005, p. 78). Interestingly, the tourism development plan was not proposed by someone from the tourism industry or government agencies, but by a professor, among other people, from the Department of Foreign Relations of the Central University of Politics, who claimed that "tourism is not merely a form of education, but a very important part of education in general" (Zhu, 1945; see Wang, 2005, p. 89). In promoting tourism to the public, the plan particularly targeted the youth in hope of persuading and recruiting them to work in the remote border regions by instilling a kind of pioneering spirit.

The normalization of organized political tours was a significant development of tourism in the Republic of China. In 1937, the Executive Yuan, the executive branch of the Central Government of the Republic of China, invited more than 180 leaders from government departments, agencies, and industries for a two-month tour along the Nanjing-Yunan Road. During the "long march," the group members did many kinds of propaganda work such as spreading "the goodwill" of the KMT in addition to visiting the touristic spots (Jia, 2004, p. 86).

The journalists' tour is worthy of a particular note due to its propaganda nature. According to a report published in *Shenbao* in 1946, a group of 73 journalists from Shanghai, claimed to be organized by a journalist association for a tour of Wuxi, was invited to the headquarter of the First Pacification Zone of the KMT immediately upon their arrival. The military provided the journalist group all kinds of luxury services for the rest of the tour including lunch, tea parties, banquets, hotels in addition to transportation. Gen. Tang Enbo, the Commander in Chief of the Zone, attended the dinner party with the journalists (Jizhe, 1946).

In Republican China, tourism, patriotism, and nationalism were intertwined. Compelling evidence points to the fact that modern tourism in China had already evolved into a propaganda apparatus during the KMT regime. On the one hand, it was a continuation of a much longer Chinese culture of tourism with an assumed function of propaganda, and on the other hand, an abrupt departure from the laissez-faire business mode of modern tourism without much government interference. Perhaps more significantly, political tourism during the KMT period resembled a prototype of Red Tourism, which was looming large on the horizon even prior to the Communist Party of China (CPC) coming to power.

"Playing politics": the propagandistic transformation of tourism, 1949–1965

The leisure business withered quickly after 1949. Yi (2009) called the first five years of the People's Republic (1949–1954) the "regression period" of China Travel Service as the company continued losing money after 1949 and finally went bankrupt in 1954. The bona fide reason for the regression was the replacement of the leisure middle class by the proletarian class in the wake of sociopolitical transformations. In the first 30 years of the New China, tourism as a sector in the nation's economy was non-existent thanks to its internal political system and external political and economic blockades imposed by the West. Sofield and Li (2011) described this period as "three decades of 'non-tourism' under Mao" (p. 503).

Tourism as a bourgeois lifestyle and an emerging industry in the KMT period could not survive in the Mao era unless it could transform into something else, something ideologically appealing to the newly established People's Republic. Tourism needed a new role, in a nutshell. In a memoir, Yang Gongsu (1994), a diplomat in charge of tourism affairs during the Cultural Revolution, recalled that in an internal meeting of the Ministry of Foreign Affairs, the head stated that "tourism work is part of diplomatic work, which is about playing politics" (p. 52). Oddly, this role of "playing politics" echoes what later President Jiang Zeming brought forward as journalism's role of "talking politics." Then what does "playing politics" mean?

In this section, I argue that the meaning of "playing politics" was twofold: domestically, it denotes a specific nation-building function of political tourism; internationally, it suggests a diplomatic role of inbound tourism. Together, tourism's role of "playing politics" suggests that a propagandistic transformation of tourism was underway.

Tourism as a tool of mobilization for nation building

1949 was a watershed in the development of Chinese tourism, dividing the "old" yet modern model of tourism from the CPC's new model. This was an ideological turn. It was ideological not only because this change occurred as a result of the ideological transformation from the Republic of China to the People's Republic but also because the new model enormously expanded the ideological dimension of traditional Chinese tourism to the extent that tourism now had become a legitimate tool of propaganda.

The Chinese tourism bureau operated as a government agency since 1949. It specifically dealt with matters that were thought to be critical to foreign relations. In virtue of this, tourism work was politics-oriented (Duan, 2002). In an institutional reform in 1958, the Secretary-General of the State Council was directly in charge of the China International Travel Service (CITS), and accordingly, all CITS's subordinate branches were handed over to local governments (Han, 2003). Chinese scholars have commonly characterized PRC's tourism prior to the 3rd Plenary Session of the 11th Central Committee as a "accommodating model." Whether referred to as "accommodating for political purposes" (Wang & Wang, 2001), "accommodating by government" (Gao, 2006), "accommodating for diplomatic purposes" (Li, 2010), or even a "malpractice" (Duan, 2002), this model of tourism was political.

The new model mushroomed in the absence of the regular tourism. It generally showcased the progress of socialist China. Yi (2009) has noted that organized political tours such as the "production study tour" (*shengchan guanmo tuan* 生产观摩团), or the "industry and commerce study tour" (*gongshang kaocha tuan* 工商考察团) became China Travel Service's core business in the first years of the PRC. Correspondingly, the scope of *China Traveler*, the biggest travel magazine at the time, changed dramatically from aiming at promoting the tourism business to spreading culture and education with an emphasis on indoctrination of patriotism. In light of this emerging educational function, the travel magazine became a textbook for some schools, and its subscribers changed from the educated middle class in the KMT era to organizational subscribers, mainly schools and military troops. The paid circulation snowballed from 8,200 to 19,000 copies a year after the Shanghai Post Office took over the magazine's distribution in 1953 (Yi, 2009).

Tourism became a part of government-organized celebrations. To celebrate the first Children's Day after 1949, the Beijing government opened parks, scenic spots, and historical sites for children to visit in addition to a series of commemorative meetings (Pingshi choubei, 1949). Not all tours for celebration purposes were free. On May 1, 1949, the International Labor Day, the Confucius Temple and the Temple of Heaven in Beijing reopened to the public for a five-yuan admission fee (Piengshi datan, 1949). On the New Year's Day 1950, all Beijing touristic spots opened and admission was free for troops, relatives of revolutionary martyrs and military families, and half-price for all others (Bai, 1949). Note that in the Mao era, people organized by their units (*danwei* 单位) sometimes visited historical sites, now "Red" spots, for various political purposes. Nevertheless, the tourists did not want to be called "tourists" at the time (Gao, 2006). This is another manifestation of the ineradicable anti-leisure ideology of tourism rooted deeply in Chinese culture.

Throughout the 1950s, the *People's Daily* published a series of photos featuring well-known Chinese tourist spots on the left of the newspaper's masthead under the running title "Our Great Motherland." It is very rare in the history of world journalism for a daily newspaper to constantly place tourism photos in the most eye-catching position for a long period of time. This demonstrates the party organ's intention of instilling a well-crafted, themed visual representation of the Chinese nation-state in the public by which the new nation was imagined.

Minority groups were critical to the new nation building. The Chinese state proclaimed that the Chinese nation consists of 55 ethnic minority groups in addition to the Han Chinese majority. But this was rather a matter of rhetoric than an already-accepted social construct in early socialist China. In other words, the image of a 56-ethnic groups-based "Great Motherland" needed to be propagated. And political tourism was an ideal vehicle for that. The Chinese government dispatched delegations to minority regions right after the establishment of the PRC. Similar to what I have remarked on as the goodwill mission of the Nanjing government during the KMT period, the CPC delegations aimed at propagandizing the party's new minority policy. In 1950, the Central Government decided to dispatch the first mission to southwestern China where many ethnic groups lived together and accordingly, ethnic relations were complicated. More than 120 members from government departments and agencies were given a month-long training about histories, habits, and customs of the ethnic groups prior to their departure (Zhong, 2013). In a front-page editorial of the *People's Daily*, the mission was said to be a "true symbol of friendship and cooperation among all ethnic groups" (Song, 1950). In return, minority elites were invited to Beijing for political tours, which became a "new tradition" of Chinese political tourism (Bulag, 2012, p. 137). In 1950, minority delegations divided on a regional basis were invited by the state to join the national celebration of National Day (October 1) held in Beijing (Xinjiang, 1950). The *People's Daily* reported the event in a full page of news photos under the headline: "Let All Ethnic Groups of the People's Republic of China Unite." The introductory text reads:

> Representatives of brother ethnic groups of the People's Republic of China came to Beijing from remote regions to celebrate their own country's birthday. They represent thirty-six ethnicities in our country, and this is the first ever grand reunion of all brother ethnic groups throughout Chinese history, a guarantee of building the national unity of the People's Republic of China.
>
> (Ge, 1950)

Tourism as a diplomatic tool

"Playing politics" was consistent with Premier Zhou Enlai's instructions on Chinese inbound tourism. Zhou described the goal of tourism work was "to *xuanchuan* ourselves, understand others, wield influence, and gain sympathy" (Yang, 1994, p. 52). That is, for the young People's Republic, inbound tourism was a diplomatic tool.

In 1954, the state-owned tourism company "China International Travel Service" (CITS) was established. It was the same year that the privately owned China Travel Service closed down. The word "international" that distinguishes the name of the new company from that of the old one has a specific meaning: "diplomatic." Situated in the context of citizen diplomacy, *minjian waijiao* 民间外交 (aka people's diplomacy), CITS served as a diplomatic agency for China to communicate with the outside world. One of major tasks of CITS was to provide travel services to visiting overseas Chinese, the tourists whom the party believed to have potentials to spread out the positive image of the PRC to the rest of the world in addition to serving the country directly or providing economic aid (Quanguo, 1956). In 1957, CITS created its subsidiary company, Overseas Chinese Travel Service, with 35 branches scattered throughout the country. During an eight-day conference held in March 1957 in Beijing, delegates of the Overseas Chinese Travel Service reached a consensus that unlike for-profit corporations, the agency is not for profit and as such, all the profits will go back to the tourists (Bianli, 1957). Not coincidently, Xiamen Overseas Service, the first tourism company born in the PRC, also exclusively targeted overseas Chinese (Han, 2003).

1964 was decisive for the development of the PRC's international relations. It turned out that the same year was also of historical significance to Chinese tourism. Premier Zhou Enlai visited 13 Asian and African countries in February. During the tour, Premier Zhou exchanged views and reached extensive agreement with the Third World leaders on a wide range of world issues such as expansionism and imperialism. The event led to a wave of institutionalization of international relations in China as Zhou Enlai ordered the establishments of several universities and departments to train and prepare diplomatic staff (Shambaugh, 2002). In July, the Standing Committee of the National People's Congress (NPC) approved the establishment of the China Travel and Tourism Enterprise Administration (now the CNTA) under the State Council. Although named "enterprise," it designated a special practice in this particular socialist context as shown in government documents: to run a business without considering cost and market rules (Gao, 2006). Again, what really mattered to the tourism agency was not only profit but also politics.

Established in December, the Tourism Administration had two specific goals: (1) politically, to propagandize its social/socialist achievements worldwide, and (2) economically, to accumulate foreign exchange from foreign tourists (Han, 2003; Yang, 1994). Both aimed at inbound tourism, for domestic tourism as an industry was yet to be developed. Because of its focus on international relations, it is not surprising that the tourism agency operated under the Ministry of Foreign Affairs during the Cultural Revolution and after that until the early 1980s (Han, 2003; Yang, 1994).

In 1957, the first group of tourists at their own expense toured the Soviet Union. According to a series of reports, the Chinese tourists arrived in Moscow prior to the new year eve of 1957. Although the tourists paid for everything themselves, the overseas tour was overtly political. They visited factories and the subway in Moscow and historical sites associated with the Russian Revolution, the Workers Cultural Palace, and the Russian cruiser *Aurora* in Leningrad (St. Petersburg) (Woguo shoupi, 1957; Woguo zifei, 1957).

More than half a century later in July 2015, Beijing and Moscow jointly launched a similar tourist route named "Red Circuit" in the wake of the surging Red Tourism wave. The Red Circuit features the life of Vladimir Lenin and the Russian Revolution (Russia, 2015). Unsurprisingly, Moscow and St. Petersburg were both included in the four Russian destinations of the Red Circuit.

Coincidently, the diplomatic scope of tourism echoed Mao's personal view of tourism as a means of making friends, particularly those who could help achieve his political goals (Ling, 1994). For Mao, tourism was a very effective tool for political communication. This was, in fact, a prevailing perspective of tourism among early Chinese thinkers. Cai (1994) has remarked that Mao's attitude towards tourism was influenced by ancient Chinese thinkers. Like Confucius, Mao opposed touring exclusively for leisure. In Mao's class notes it reads, "a tourist should not only enjoy the scenery, but also make friends with prestigious figures, learned intellectuals" (Sun et al., 2008, p. 300). During his first sojourn in Beijing, Mao furthered friendship with the leaders of the New Culture Movement, including Li Dazhao 李大钊 (1889–1927), Chen Duxiu 陈独秀 (1879–1942), Cai Yuanpei 蔡元培 (1868–1940), Tao Menghe 陶孟和 (1887–1960), and Hu Shi (Cai, 1994).

The perspective of treating tourism as a valuable source of foreign exchange was criticized during the Cultural Revolution, rendering Chinese tourism completely political at that time.

The Great Rally: a prelude to Red Tourism, 1966–1976

One way to see the Cultural Revolution is to view it as an assemblage of many "great" political movements such as the "Great Rally" (*da chuanlian* 大串连), the "Great Leap Forward" (*da yuejin* 大跃进), and the "Great Criticism" (*da pipan* 大批判). While the term "great" denotes massive scale, it also indicates mass movement in the true sense beyond its figurative sense. In other words, the Cultural Revolution can be an exceptional case for studying the nexus between mass communication and mass mobility, and therefore, a starting point for probing into modern Chinese mass political tourism.

Retrospectively, the Great Rally paved the way for the later emergence of Red Tourism. Zhou and Gao (2008) have framed the political movement as a special phase of Red Tourism. Likewise, Huang (2010) considered it the origins of Red Tourism. In Yan's (1993) account, the Great Rally was the "largest and most complete political tour in human history" (p. 4). It was considered *the* largest because at least 20 million people, three to five percent of the country's population, embarked on the grand political tour. The tour was purely political as it was initiated for political purposes, people joined it for political reasons, and it ended in political struggles (Yan, 1993).

The Great Rally was intended to be a "seeder" for the Cultural Revolution through human movement. The political movement was believed to be self-organized at the early stages until the *Red Flag* magazine published Mao's speech in which Mao rallied the masses to partake (Yan, 1993). The political movement transformed into political tourism when the government announced that the government would cover all travel-related expenses including train tickets, lodging, and food. Considering everything was free, Yan (1993) remarked that "it would be unwise not taking the tour" (p. 15). It has been commonly assumed that the majority of participants of the Rally were the Red Guards. But scholars have shown otherwise: the majority actually were students and teachers mainly from colleges and high schools (Yan, 1993; Zhou & Gao, 2008). Political tourism was particularly attractive to the youth since they had nothing to do when all schools were forced to be closed at the time. Their motives were varied. Some were enthusiastic about revolutionary rebellion. Still many others were rather lukewarm towards that, joining the Rally as an escape from harsh political reality.

In the meantime, a kind of prototype national "Red" tourism network emerged. In contrast to all other historical relics that were destroyed in the name of "Eradicating the Four Olds" (*po si jiu* 破四旧), the revolutionary sites were protected, enshrined, and reused. Given that millions of young pilgrims arrived each day, those sites had gained popularity almost overnight and were immediately transformed into tourist spots with free services and food and all costs borne by the local government. The most popular destinations of the Rally were Yan'an, Jinggangshan, Shaoshan, and Ruijin. Today they stand as the most popular Red Tourism cities. It should be noted further that other than historical sites, other places, particularly those remote ones such as Xinjiang and Yunnan, having less to do with the Chinese Revolution but with spectacular scenery, also turned into hot destinations during the Rally. Previously having to pay considerably, now people were able to travel to these scenic places at the government's cost.

But political tourism was not as fun as it may sound whatsoever. Given terrible transportation, poor lodging facilities, among many other difficulties, many tourists suffered badly during their journey. For some, the free political tour was even at the cost of their human dignity and life. Many witnessed that travelers were so overpacked like sardines in the train that they were unable to move within tens of hours of travel (Yan, 1993). Some suffocated to death on the train (Xia, 2004). In April 1967, the Central Committee, State Council, Central Military Commission, and the Central Cultural Revolution Group issued a notice to end the Rally nationwide.

Notwithstanding its political orientation, it is arbitrary to translate Chinese tourism policy prior to the economic reform as the "fewer outsiders the better" (Richter, 1983, p. 397). Contrarily, foreign tourists were surprisingly welcomed by the Chinese government as far as the propaganda goal was concerned. In the Mao era, it became a common practice in the party's propaganda work to report private visits made by American tourists in the *People's Daily*. A quick catalog of the reports of foreign tourists shows a wide range of professions, from lawyer, mathematician, scientist, scholar, writer, etc. For "leftist" foreign tourists, mainly foreign communists or those in sympathy with the CPC, travel expenses were paid by the Chinese government (Yang, 1994). In 1971, when China was admitted into the United Nations, Mao instructed the tourism agency, "the number (of foreign tourists) should be increased and include rightists as well" (Yang, 1994, p. 52). According to Yang Gongsu, the head of the tourism agency at the time, "rightists" referred to government officials, journalists, and businessmen from Western capitalist countries. This practice seems to reconfirm what Mao had said to American diplomat John Service many years ago during the Chinese Civil War that the presence of American travelers itself is good propaganda.

The economic turn, 1977–1989

The starting point for the development of modern tourism in the PRC is marked by a series of speeches given by Deng Xiaoping from 1978 to 1979. Du Yili (2012), Deputy Director of CNTA, described the situation of the tourism industry prior to Deng's leadership as the "Five-No State," namely, "no concept, no market, no scale, no goal, and no policy" (p. 57). Simply put, no tourism industry whatsoever.

Against the backdrop of economic reforms in the 1980s, the rapid expansion of Chinese tourism was strategically aimed at inbound tourism while starting up domestic tourism. Han (2003) regarded this inbound-tourism-based tourism development plan as "the Chinese model of tourism," a kind of "Chinese characteristic" in the realm of tourism (p. 24). Starting from the late 1970s, tourism in China switched swiftly from its diplomatic role to an economic one.

From 1978 to 1979, Deng Xiaoping gave five speeches particularly on tourism, all of which provided guidelines for the development of tourism in the post-Mao era. Duan (2002) pointed out that these talks answered two critical questions about why and how to develop tourism in China. In light of the economic reform, Deng believed that the development of tourism could bring about breakthroughs in emancipation of the mind, internal reform, and opening up (Duan, 2002; Zhang & Xie, 2005). In contrast to the diplomatic orientation of early PRC tourism, Deng valorized the economic role of tourism as an industry to boost the nation's economy. In October 1978, Deng gave a speech to the heads of the Civil Aviation Administration and the Tourism Administration after meeting Pan American World Airways CEO William Seawell. Deng pointed out that for China, the largest source of tourists was the United States. Deng calculated,

> if each foreign tourist would spend $1,000 in China with a total of 10 million of tourists visiting China a year, China could earn $10 billion, and even if only half of the tourist number, it is still $5 billion.
>
> (Deng, 2000, p.2)

In view of the lack of proper diplomatic channels, Deng instructed the two government agencies to get into the international tourism market through private channels. In a talk with the head of the Tourism Administration in 1979, Deng rearticulated his economic orientation, saying that the reason to develop tourism was to increase revenue, making every effort to make it profitable (Deng, 2000). Other socialist countries did not have a similar development pattern as China to make tourism what Sofield and Li (2011) define as a keystone industry.

At the heart of what Chinese scholars have termed "Deng Xiaoping's thought on tourism" was the money-making ideology of the market economy. Propaganda as a political agenda went unmentioned in both Deng's and later Chinese scholars' accounts of tourism in this particular period of time. This is, perhaps, in part due to the depoliticization trend in the aftermath of the Cultural Revolution. Nevertheless, on June 9, 1989, five days after the Tiananmen incident, Deng

Xiaoping admitted that the biggest mistake being made during the ten-year reform was not drawing much attention to ideological and political education, which he called a "serious error" (Deng, 1994). And this was critical to the coming of Red Tourism.

The making of Red Tourism, post-1989

Given the nature of the Cultural Revolution, it is not surprising that tourism was entirely political as everything else. What is surprising, though, is the party's resumption of the political tradition of Chinese tourism in the 1990s to launch, institutionalize, and promote Red Tourism after tourism had been depoliticized and become a pillar industry in Deng's era. Note that Confucianism alongside a wholesale rhetoric about cultural heritage was also widespread about the same time. This is by no means a coincidence.

The ideological vacuum

It took about 20 years for China to switch its cultural heritage politics from one extreme to another. In 1966, in light of the "Eradicating Four Olds" (old customs, old culture, old habits, and old ideas) movement at the beginning of the Cultural Revolution, cultural heritage was generally deemed as counter-revolutionary. In 2006, cultural heritage became something that the government encouraged the whole nation to celebrate on Cultural Heritage Day. The state politics towards Confucianism was just the same. Once being criticized throughout the Cultural Revolution, Confucianism started being institutionalized in 2004 as Confucius Institutes grew worldwide with support from the state. At the intersection of cultural heritage and Confucianism is heritage tourism. And at the heart of heritage tourism is propaganda/*xuanchuan*. In that sense, Blumenfield and Silverman (2013) see cultural heritage as a "political tool" for "those in positions of power telling stories about the past and present" (p. 4). Perhaps, a propaganda tool would be more precise, particularly considering that they figuratively describe heritage tourism as the "genie let out of the bottle" (Blumenfield & Silverman, 2013, p. 9).

Inquiring into the recent history of Red Tourism, I argue that the rise of Red Tourism can be seen as a strategic response to what I would call an "ideological vacuum" in the aftermath of 1989. The loss of credibility of major revolution ideologies in Mao's era – Communism, Maoism, and Leninism – and the abandonment of Western liberalism in the wake of the Tiananmen incident of 1989 have created an ideological vacuum that led to a legitimation crisis for the Party. Chen (1995) defined the crisis as "three belief crises," specifically, the Chinese losing faith in the Party, socialism, and Marxism. But I have reservations about the Marxism part.

Siding with Eagleton (2011), I consider Marxism more of a critique of capitalism than a revolutionary ideology. As such, although the public might not talk much about Marxism since the end of the Cultural Revolution, it is still very hard to imagine a real "crisis" of Marxism occurring during the reform and opening period. It would be amazing, if the Chinese had lost their faith in Marxism, and alternatively, wholeheartedly embraced capitalism. If this had been the case, China would have already lost its "characteristics," and consequently, what I am discussing here, "Red Tourism," would have been, more appropriately, renamed as "Dark Tourism." Paradoxically, Marx's more stimulating ideas, such as the "alienation" of stratified human society and the "commodification" of social life, have shed additional light on China's sharp increase in social inequality occurring after the dismantling of its system of centrally planned socialism.

In reality, the pursuit of Chinese characteristics together with the ideological vacuum have left room for a reconsideration of the ideological foundation for a post-reform China. Instead of looking into Marxism, this time the party looked further back to its own history and cultural traditions. Patriotism then became an ideological magnet.

Re-signification of patriotism

In view of the "serious error" in ideological education, the CPC launched the "Patriotic Education Campaign" in 1991. It signaled the start of the making of Red Tourism by wedding "Red" to "Tourism." It was more than a resumption of the patriotism tradition of Chinese political tourism; it was a re-working of a mass communication network that involved reordering its superstructure and constructing a new base.

Central to the massive nationwide campaign, however, was a humiliating history of a colonial, semi-colonial, and semi-feudal China charged with the Chinese Revolution that mobilized the youth. Called "patriotic education," the propaganda campaign unmistakably echoes "citizenship education" in US history (Gordon, 1971). According to an "Outline on Implementing Patriotic Education" issued in 1994, patriotism now was elevated as the "common spiritual pillar for the people of all ethnicities," the "foundation of constructing socialist spiritual civilization," the "theme of our society," and the "guideline for ideological education" (Aiguo, 1994). The major locus of the ideological campaign was the education sector. Schools' textbooks were rewritten, replacing the class-struggle narrative with a new one showcasing the CPC's unparalleled role in ending the one-hundred-year humiliating history of the Chinese nation (Wang, 2008). Besides classrooms, the Outline of 1994 specifies that "Patriotic Education Bases," mainly those "pre"-Red Tourism sites, are the loci for implementing patriotic education. In 1991, the Central Propaganda Department (CPD), the Ministry of Education (MoE), the Ministry of Culture (MoC), the Ministry of Civil Affairs (MCA), the Central Communist Youth League, and the State Administration of Cultural Heritage (SACH) issued a "Notice about Conducting Education of Patriotism and Revolutionary Traditions through the Use of Historical Relics." The document requires all schools, from primary school to college, to take students to the museums and historical sites significant to the Chinese Revolution for patriotic education and to treat such tours as a must-attend educational activity (Central Propaganda Department (CPD), 1991). The 2004 Outline reiterates the key role of the bases/historical relics and museums in ideological education, calling for free admission to all school-organized tours of these bases (CPCCC & the State Council, 2004). About that time, Red Tourism surfaced.

In 1997, the Central Committee of the CPC (CCCPC) compiled a list of 110 National Exemplary Patriotic Education Bases that further promoted the expansion of such sites (Zhou & Gao, 2008). By the end of 2005, the total number of National Bases reached 270 in addition to more than 10,000 Bases at the provincial level (National Red Tourism Work Coordination Group, 2008).

Tourism in post-socialist China forged a double identity: an industry on the one hand, and a social force on the other. While the economic development of China's tourism after 1978 has been stressed by tourism practitioners and scholars, the second identity of Chinese tourism as a social force in social development can never be overstated. In light of the bureaucratic structure of macroeconomic regulation and control, tourism is under the control of the Social Development Division of the National Development and Reform Commission (NDRC), a supra-ministerial cabinet agency under the State Council. In addition to tourism, the Social Division also has administrative and planning control over other cultural industries including sports, television, radio, news, and publishing. This is to suggest that the party sees all these cultural industries as special ones that would bring about critical changes in the society in addition to economic growth.

The social function of tourism in China can be better understood as social integration, a main characteristic of *xuanchuan* that I have elaborated in Chapter 1. In the lexicon of the contemporary CPC's propaganda work, tourism in post-socialist China has been oftentimes associated with the "Construction of Spiritual Civilization," a propaganda campaign launched in 1996 (Chen, 1997; Gao, 2006; Lin, 2011). Regarding tourism as a form of mental activity, for example, Yuan (1998) has argued that tourism can work on people's mentality by improving citizenship, by facilitating communication across cultural boundaries, and by promoting cultural heritage. Gao (2006) noted that tourism can significantly contribute to the construction of a harmonious socialist society for it is a carrier of good spirit.

The "government-led model" of tourism was misinterpreted as the development model of tourism with Chinese characteristics due to the planned economy tone in its calling. Ironically, the term was initially coined by CNTA in 1996 to describe the decisive role of government in tourism development in two specific foreign countries, Israel and Turkey, after their visits (Gao, 2006). In other words, government involvement in tourism development is not unique to China, and by no means determined by a given economic system.

The coming of Red Tourism

The coming of Red Tourism has been grounded in the context of the economic departure from a centrally controlled economy to a market economy, and an ideological continuation rooted in the prior form of socialism. The structural transformation of the nation's economy resulted in a protracted and intensive round of capitalist-style privatization, which in turn has rendered strict adherence to the socialist ideology less promising in the private sector. The CPC sees the first part of this complicated process as the "construction of socialist material civilization" and the second part the "construction of socialist spiritual civilization." The development of Chinese society was said to demand social equilibrium between the two kinds of construction. This idea is epitomized in Deng Xiaoping's famous saying, "seize with both hands, both hands must be strong," which became a general guideline for the construction of socialism with Chinese characteristics.

Surprisingly, as a business model Red Tourism did not start from well-known revolutionary sites but from some unknown towns and villages around the mid-1990s. It is not so surprising, however, given the fact that the Chinese economic reform was also initially introduced in small villages. Both indicate that their attempts were so bold that they could only be made in places with less political constraints and more freedom in embracing a market economy. Based on CNKI (Chinese National Knowledge Infrastructure) databases, the appearance of the Chinese term "*hongse lvyou* 红色旅游" (Red Tourism) in mainstream publications can be traced back to 1996 in an article entitled, "Deep feelings about the Old Region" (Zheng, 1996). An interview of the head of the Women's Committee of China's Old Region Development Promotion remarks that three small towns in the old revolutionary base areas of Hebei, Shandong, and Hubei provinces conducted Red Tourism with help from a travel agency run by the Ministry of Water Resources in an effort to alleviate local poverty. In this case, the looming Red Tourism business was well in concert with the emergence of township and village enterprises (TVEs) in that both aimed to mobilize local resources for local economic development. Additionally, both changed their early trajectories by virtue of a seemingly ceaseless round of capitalization. I will illustrate this point in the last chapter on the political economy of Red Tourism.

Besides the unprecedented expansion in the Bases, there were two other propelling forces behind the rise of what Gao (2006) called the "big wave" of Red Tourism in the late 1990s. The first was that tourism had grown into a profitable sector and an economic pillar contributing greatly to national economic growth. The second was that the party's commemorations and celebrations related to its revolutionary past settled into a tradition (Gao, 2006; Huang, 2010). Gao (2006) further suggested that the anniversary significant to the CPC's history can even be a reliable predictor for a strong Red Tourism market. When the term "Red Tourism" first appeared in the *People's Daily* in 1999, the business had already become phenomenal. According to the report, Ruijin – a county-level city of Jiangxi Province where the Chinese Soviet Republic was established in 1931 – received more than 100,000 tourists in the first half of 1999 (Rui & Xie, 1999).

Having seen the profitability of Red Tourism, private enterprises soon joined. Where there was no "Base" immediately available to run the Red business, entrepreneurs created one from scratch. In 1998, a tourism firm called Huihai was created with registered capital RMB 3 million ($460,000) in Qionghai, a small coastal city of Hainan Province now best known as the permanent venue for Boao Forum for Asia (BFA). For some Chinese, Qionghai is known as the home of the legendary Red Detachment of Women, a revolutionary story adapted into many forms of literary and artistic works. The ballet version of the Red Detachment was performed for US President Richard Nixon on his visit to China in 1972. Finding a business opportunity behind the revolutionary heritage, Huihai began to build the Red Detachment of Women Memorial Park in 1998 with a total investment of RMB 25 million ($3.85 million). A fascinating mixture of revolution-themed sculptures, a museum, tropical coconut groves, and live performances with four former women soldiers of the Red Detachment who are over 90 years old, the first-generation Red-themed park opened in 2000. With half a million tourists arriving each year since its opening, the memorial park was soon included in the list of "Exemplary Bases" by the CPD in 2001 and recognized as a "National Classic Red Tourism Attraction" by the NDRC, CPD, and CNTA in 2005. During this time, the CPD allocated RMB 3 million to establish the Red Detachment of Women Memorial at the park.

First red, last tourism

In early 2004, the CPD passed on an important instruction entitled, "To actively develop Red Tourism," from Politburo Standing Committee Member Li Changchun to the CNTA. While its consequences are profound, three direct ramifications were clear. First, Red Tourism immediately became a central topic of the National Tourism Conference held in May, resulting in the "Declaration of Zhengzhou" signed by six provinces and municipalities including Beijing and Shanghai. Calling for "branding Red Tourism, raising Red Tourism to climax," the Declaration is a set of guidelines for the participating members to share resources and information, to co-develop markets, and to seek common interests in running the Red business. It proposes a network of Red Tourism routes that start at Shanghai, run through southern and northern China, and end at Beijing. It demonstrates the members' confidence in the Red Tourism market on the one hand and their burning ambition to seize monopolistic power in the early development of that market on the other. A follow-up commentary in the *People's Daily* states:

> Red tourism is both a conceptual innovation and an industrial innovation. The development of Red Tourism will help explore a virtuous cycle in which, under the socialist market economy, social benefits and economic benefits are well integrated, spiritual wealth is transformed into social wealth. Together, they are of benefit to the society and the people.
>
> (Wang et al., 2004)

In August, the *People's Daily* published a series of seven reports on the development of Red Tourism in different locations under the running title of "Red Tourism Series" in seven consecutive days. The grand gesture of the party's organ had clearly signified a new chapter of Red Tourism.

The second ramification was the "2004–2010 National Red Tourism Development Outline" issued by the Central Government at the end of 2004. It is considered the general guideline for subsequent Red Tourism developments (Zhou & Gao, 2008). In the document, the CPC provides an official definition of the term for the first time, which is:

> Red tourism mainly refers to the thematic tour for cherishing memory and learning in those memorials, historical sites, and markers that as carriers, contain revolutionary history, revolutionary stories, and revolutionary spirits and represent great achievements of the Chinese people under the leadership of the CPC during the revolution and war periods. The development of Red Tourism has immense practical significance and far-reaching historical significance for strengthening revolutionary tradition education, enhancing the sense of patriotism of the whole nation, particularly the youth, cultivating the national spirit, and promoting socioeconomic development of the old revolutionary base areas.
>
> (CPCCC & the State Council, 2004)

The 2004 Outline articulates the guiding ideology, basic principles, and development goals in propagation of Red Tourism. It also provides a roadmap for the development of Red Tourism, showing that the Chinese state was determined to develop Red Tourism at an industry level. In view of the relationship between economic and social gains, however, the Outline states that social benefit must be prioritized. A follow-up editorial of the *People's Daily* concludes, "Red Tourism is a systematic project of the society, an invention with distinctive Chinese characteristics, and multiply significant to politics, culture, and economy" (Jiji, 2005). In the meantime, the National Red Tourism Work Coordination Group (aka the "Red Office") was established.

The members of the "Red Office" are the heads from 14 government departments and agencies (Figure 4.1). In light of the bureaucratic structure of the Coordination Group, it is surprising that CNTA, the government agency in charge of tourism directly under the State Council, is at the bottom. This can be an indicator that reconfirms what Airey and Chong (2010) have said: that national policy-making for tourism in China is fragmented with CNTA only having a "comparatively low administrative status and weak bargaining power" (p. 311). On the other hand, the fact that the CPD as one of the leading agencies and ministries ranked second reveals the propaganda nature of Red Tourism. That is, the first is Red, last Tourism.

In an interview, Susan, an official in charge of Red Tourism at a provincial-level tourism bureau nicely illustrated this point from a bureaucratic insider point of view, which I present at length below:

> The parent units of Red Tourism are the department of propaganda and the Cultural Affairs Bureau (*wenhua ju* 文化局). In fact, the Tourism Bureau cannot manage it (Red Tourism). The National Red Tourism Work Coordination Group is set up in CNTA, but the group top leader is from NDRC, not from CNTA. Only the office is located in CNTA, [whose role is] merely delivering

announcements. Like CNTA, the local Tourism Bureau is where the local Red Office is located, and still, the leader is from the local NDRC bureau. There is no specific government funding/subsidy and post in the Tourism Bureau specific for Red Tourism. The heads of our Red Office come from as many as twelve government departments with only four really in charge: the local NDRC Bureau in charge of money, the Finance Bureau for distributing money, the department of propaganda taking command, and the Tourism Bureau laboring at drafting and other specific works.

As a matter of fact, the department of propaganda does not really work on propaganda [here Susan means "advertisement" in particular]. For example, we planned to take out a full-page advertisement in the largest local newspaper for Red Tourism. According to the division of work, the department of propaganda should take on this work. But they said they cannot do this unless they have received direct instruction from the top authority. Ultimately, we had to undertake the task ourselves using funds taken out from elsewhere. The quota of funding we get from NDRC is only for fixed uses, say, 5 million yuan for advertisement, and Red Tourism is not included. Since the funding is from the government for the development of tourism in general, and the development of Red Tourism is part of that, we [think it would be fine to] move some funding for Red Tourism. But we do not have a quota particularly for Red Tourism. Except for Guizhou Province which has two established posts for the Red Office, elsewhere staff from local tourism bureaus run Red Offices as a concurrent post.

(interview, December 17, 2015)

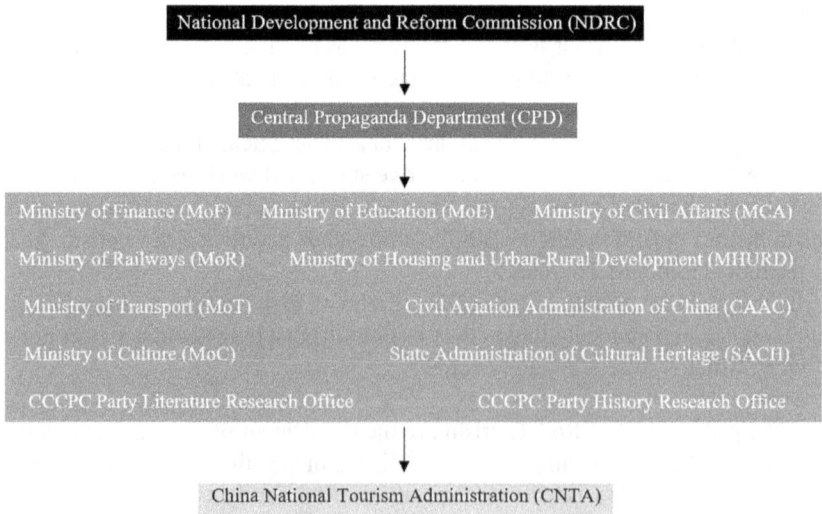

Figure 4.1 The bureaucratic structure of the "Red Office." Created by the author.

The third ramification is that, named as the "Number One Project" for its cardinal importance, rebuilding the image of three major patriotic education bases, namely, Shaoshan, Jinggangshan, and Yan'an, started in 2004. Each place was branded with a specific selling point, Shaoshan for "Mao Zedong leading the Chinese people to stand up," Jinggangshan, "the Chinese revolution sparking from here," and Yan'an, "Yan'an Spirit and the sacred place of the revolution" (Zhou & Gao, 2008, p. 8). By now, previously revolutionary sites had completed their two-step transformation, first to patriotic education bases, then Red Tourism sites. Accordingly, Red Tourism also transformed from an undefined kind of touristic activity to an institutionalized industry with specified superstructure and regular bases in addition to full support from the state. In the following three years, the Central Government allocated more than RMB 1.5 billion ($231 million) for building Red Tourism sites and museums and the total investments from local governments were even more than that number (National Red Tourism Work Coordination Group, 2008, p. 31). According to the National Bureau of Statistics, in 2004 China's Red Tourism sites received 120 million visitors, counting for 10 percent of tourists in the country and the consolidated revenue of Red Tourism reached RMB 20 billion ($3.07 billion) (Zhou & Gao, 2008).

In 2011, these numbers were multiplied with 540 million visitors counting for 20 percent of total domestic tourists (Jin, 2012). In the same year, an upgraded version of the 2004 Outline was issued. With an industry of Red Tourism already becoming reality, the 2011 Outline substantially upgraded the role of Red Tourism from previously promoting patriotic education to fostering the socialist core value system. In the "Guideline" section, the Outline of 2011 states:

> As a political project and a cultural project, Red Tourism must emphasize its pivotal role in facilitating the construction of the socialist core value system and imbuing the masses with the idea that it was the history and the people that chose the CPC, the socialist system, and the road to open up and reform. In doing so, Red Tourism will help reinforce public trust in the CPC and in socialism with Chinese characteristics and thus consolidate the common ideological foundation for both the Party and people of all ethnicities.
>
> (CPCC & the State Council, 2011)

An article published in the CCCPC's *Qiu Shi* magazine defines Red Tourism as a "form," a "resource," a "carrier," and a "pathway" to the implementation of the socialist core value system (Deng & Gao, 2014, p. 92). Thus far, Red Tourism has been politically and economically phenomenal and phenomenally successful in terms of economic gains and popularity. According to a report, in 2015 the major Red Tourism cities completed a total investment of RMB 42.37 billion ($6.52 billion) and received 452 million tourists with the consolidated revenue RMB 286.98 billion ($44.15 billion), an increase of 13 percent (Dou, 2016).

Two problems must be addressed in interpreting these numbers. First, it should be clear that Red Tourism is profitable only in a handful of places rich in revolutionary resources, particularly those old revolutionary base areas. Red Tourism in most other areas – as it has spread out all over the country already – solely relies on government subsidies. Given Red Tourism's subjectivity, this is quite understandable. Other forms of tourism are the same. For example, seaside tourism can only thrive in coastal areas and islands, whereas the wildlife tourism industry greatly depends on wildlife sanctuaries. Liz, a head of a municipal tourism bureau disclosed:

> The money we receive from Red Tourism is totally negligible compared to how much we get from regular tourism because tourists visit all [Red Tourism related] museums and sites for free. Even if we charge a 10-yuan admission fee for the museum with 2 million tourists arriving each year, we still cannot make ends meet considering that we carry out major renovations at three-to-five-year intervals with each costing tens of millions.
>
> (interview, July 10, 2014)

The unbalanced nature of the Red Tourism business, in turn, suggests an intense profit-concentration. In view of this, Red Tourism can be a goldmine for profit chasers. This is why Red Tourism has been rapidly commodified and capitalized, changing the geography of social space of some places, whereas in other places, the survival of Red Tourism heavily depends on the local government.

The second problem is about the reliability of these numbers. The government released statistics on Red Tourism, as published in popular media such as newspapers, the Internet, and social media posts, and also cited in this study. They were basically drawn from self-report surveys. The Red Office requires all its subordinate bureaus to report Red Tourism revenues, tourist numbers, etc., and then it puts all collected numbers together to map out the national Red Tourism. Therefore, the statistics about Red Tourism as presented in this book, to a large extent, are merely indicative of the scale and the growth of the industry. Beyond this, the numbers would be less valid. Susan acknowledged:

> The national Red Office has a platform for collecting data, including the tourist numbers to Red Tourism attractions and Red Tourism revenues. Every province has a local Red Office, which is asked to report those numbers. Perhaps this is how its statistics were produced. Do you ever know where the consolidated revenue of national tourism comes from? This is very difficult to define, let alone that of Red Tourism. Take Lushan for instance. You may count everything, from meals, hotels, transportation to entertainment and touring into Red Tourism revenue.
>
> (interview, December 17, 2016)

Conclusion

Political tourism in China is writ large. While political tourism is a form or a dimension of tourism in other parts of the world, it carries the identity for Chinese tourism throughout history. In light of unfriendly attitudes towards commercial culture, Yan (1993) has argued that the origin of tourism in China was deeply political. The political orientation and tradition of Chinese tourism was rearticulated and emphasized right after 1949, despite the fact that tourism as an industry had already germinated in the KMT regime. In analyzing Chinese tourism policy, other scholars have also concluded that the development of tourism in China was politically motivated (Sofield & Li, 2011; Richter, 1983). Airey and Chong (2010) viewed tourism in Mao's era as merely a "political and diplomatic vehicle" (p. 299).

And it is mass political tourism. It calls for *all* citizens to engage and is by no means a special interest tourism (SIT) that caters to the interests of specific groups and individuals. The call is guaranteed by a form of bureaucratic-authoritarianism embodied in a systematic hierarchical structure created by many important government departments and agencies that normally are not responsible for tourism in other countries. Ideological orientations of the Red tourists are varied. They are regular tourists whom you can see in other popular tourist destinations too such as Beijing, Shanghai, Bangkok, Oahu, and Los Angeles.

Mass political tourism in China plays a role of social integration. It is a shared characteristic of *xuanchuan* that I have discussed earlier. In that sense, mass political tourism is indeed a system of propaganda. The social integration role of Chinese political tourism has been discussed by few scholars but in different terms. In a historical study of the state-orchestrated ethnic minority tours in the Mao era, for example, Bulag (2012) has conceptualized the propaganda role of political tourism as "centripetalism," the idea to integrate the marginalized ethnic groups from the cultural periphery into the center by taking ethnic minority elites to the tours. During the touring, intended ideologies were indoctrinated in a rather subtle way such as showcasing economic achievements of the New China. Grounded in the Foucauldian notion of the power of seeing/gazing, the propaganda nature of political tourism was described as "the control or management of the state image" (Bulag, 2012, p. 136).

Again, political tourism in this context carries a *neutral* connotation of propaganda. In their study of Red Tourism, Zuo et al. (2016) referred to the integrating function of political tourism as "political socialization." They argued that Red tourists may learn values and attitudes intended by the state to support its political system. Examining ethnic tourism in Tibetan China, Hillman (2009) concluded that formerly viewed as a social force for consolidating national unity, this form of political tourism "provide[s] part of the solution to changing public opinion" (p. 6). Traditionally parsed as manipulation of public opinion, now propaganda in this description rather had a desired and positive effect in that "greater understanding and sensitivity" between Tibetans and Han Chinese was achieved (Hillman, 2009, p. 6). This is an example of good propaganda.

A prevailing type of mass political tourism in contemporary China, Red Tourism is the result of a series of structural transformations of tourism throughout Chinese history. Deeply rooted in the Chinese philosophy of a "serious" tourism as manifested in Confucianism, Red Tourism represents the education orientation of traditional Chinese tourism. When tourism began to form an industry in the

1920s in Republican China, the education legacy of tourism was carried out with an articulation of patriotism. With an industry of tourism fading away in early socialist China, only its politics survived. The three decades of the political transformation of Chinese tourism has left an inerasable "Red" mark on Chinese mass political tourism. The economic transformation of Chinese tourism in the early economic reform period turned out to be a bittersweet experience for the CPC. On the one hand, tourism grew rapidly into a keynote industry. On the other hand, the effort to erase the inerasable failed dismally from the viewpoint of the Party as it believed the lack of ideological education was partly responsible for the 1989 incident. In view of this, an ideological transformation of Chinese tourism in the name of patriotic education occurred right after 1989, leading to the emergence of Red Tourism. It can be understood as a profound U-turn to the patriotism tradition of Chinese tourism in the long Chinese history across imperial, republican, and socialist China on the one hand and a reorientation of market to a market economy with "Chinese characteristics" on the other.

Now it is clear that the driving force behind Red Tourism is not a single "ism," but a mixture of many isms: Confucianism, patriotism, socialism, authoritarianism, among others. Together, it might be called what Wang (1944) termed some 70 years ago as "propagandism (*sic*)," or put another way, *xuanchuan*.

Red Tourism points to a paradigm of mass communication where space alongside spatial practice produces meanings that matter to the nation-state. What has been addressed a lot so far was the "Red" part of Red Tourism, the political articulation in the ideological dimension. What has been both lost in the government documents and my current account is tourism itself, the fun part. I will examine the social space of Red Tourism in the next chapter to explore how the fun was produced and what the fun is producing in return.

References

Aiguo zhuyi jiaoyu shishi gangyao [Outline on implementing patriotic education]. (1994). *Xuexi yu shijian, 126*, 3–8.

Airey, D., & Chong, K. (2010). National policy-makers for tourism in China. *Annals of Tourism Research, 37*(2), 295–314.

Bianli Huaqiao huiguo guanguang tanqin he canguan youlan [Convenience for overseas Chinese to return home for sightseeing and family visit]. (1957, April 3). *The People's Daily*, p. 4.

Blumenfield, T., & Silverman, H. (Eds.). (2013). *Cultural Heritage Politics in China.* Springer.

Bulag, U. E. (2012). Seeing like a minority: Political tourism and the struggle for recognition in China. *Journal of Current Chinese Affairs, 41*(4), 133–158.

Cai, L. (1994). Mao Zedong zaoqi lvyou sixiang chutan [An exploration of Mao Zedong's thought on tourism in his early years]. *Mao Zedong sixiang luntan, 2*, 74–75.

Central Propaganda Department. (1991). *Announcement about conducting education of patriotism and revolutionary traditions through the use of historical relics.* http://www.cpll.cn/law2486.shtml

Chen, G. (1997). Qianlun fengjing youlanqu zai woguo jingshen wenming jianshe zhong de zhongyao diwei [On the important role of touristic spots in the construction of spiritual civilization]. *Lvyou kexue*, *2*, 22–24.

Chen, J. (1995). The impact of reform on the Party and ideology in China. *Journal of Contemporary China*, *4*(9), 22–34.

Communist Party of China Central Committee (CPCCC) & the State Council. (2004). *2004–2010 national Red Tourism development outline.*

CPCCC & the State Council. (2011). *2011–2015 national Red Tourism development outline.*

Deng, X. (1994). Zai jiejian shoudu jieyan budui jun yishang ganbu shi de jianghua [Address to officers at the rank of general and above in command of the troops enforcing martial law in Beijing (June 9, 1989)]. *Deng Xiaoping wenxuan, vol. 3*. Renmin chubanshe. The English translation is available at https://dengxiaopingworks.wordpress.com/2013/03 /18/address-to-officers-at-the-rank-of-general-and-above-in-command-of-the-troops -enforcing-martial-law-in-beijing

Deng, X. (2000). *Deng Xiaoping lun lvyou* [Deng Xiaoping on tourism]. Zhongyang wenxian chubanshe.

Deng, Y., & Gao, J. (2014). Shehui zhuyi hexin jiazhiguan shiyu zhong de hongse lvyou chuangxin fazhan yanjiu [A study of the innovative development of Red Tourism in light of the socialist core value system]. *Qiu Shi*, *8*, 92–96.

Dou, W. (2016, February 15). 2015 nian hongse lvyou chengji kexi yuanman shouguan [2015 Red Tourism ends successfully]. *Zhongguo lvyou bao*. http://www.ctnews.com .cn/zglyb/html/2016-02/15/content_122487.htm?div=-1

Du, Y. (2012). Tiwu Deng Xiaoping lun lvyou de zhanlue sikao [Understanding the strategic thinking of Deng Xiaoping on tourism]. *Dang de wenxian*, *6*, 57–61.

Duan, Q. (2002). Deng Xiaoping lvyou jingji sixiang yu dangdai zhongguo lvyou jingji de fazhan [Deng Xiaoping's thoughts on tourism economy and the development of tourism economy in contemporary China]. *Dangdai zhongguo yanjiu*, *9*(4), 68–74.

Eagleton, T. (2011). *Why Marx was right*. Yale University Press.

Fan, Y., Zhang, X., & Wang, Y. (2008). Kongzi lvyou sixiang: yanjiu pingshu yu zai shenshi [Confucius' thoughts on tourism: A literature review and rethinking]. *Lvyou xuekan*, *23*(12), 77–81.

Feng, S. (2007). A study on Wang Mang's jade tablet for the Feng-Shan Ceremony of the Xin dynasty. *Chinese Archaeology*, *7*(1), 163–169.

Gao, S. (2006). *Zhongguo lvyou chanye zhengce yanjiu* [China's tourism industry policy studies]. Zhongguo lvyou chubanshe.

Ge minzu renmin tuanjie qilai [Let all ethnic groups of the People's Republic of China unite]. (1950, October 14). *The People's Daily*, p. 5.

Gordon, G. N. (1971). *Persuasion: The theory and practice of manipulative communication*. Hastings House.

Gross, M. (2011). Flights of fancy from a sedan chair: Marketing tourism in Republican China, 1927–1937. *Twentieth-Century China*, *36*(2), 119–147.

Han, K. (2003). *Xinshiji de zhongguo lvyouye* [China's tourism industry in the new century]. Zhongguo lvyou chubanshe.

Hillman, B. (2009). Ethnic tourism and ethnic politics in Tibetan China. *Harvard Asia Pacific Review*, *10*(1), 3–6.

Hu, Z. Y. (2009). *A study of Red Tourism in China: Exploring the interface between national identity construction and tourist experience* [Unpublished doctoral dissertation]. University of Hong Kong.

Huang, X. (2010). *Hongse lvyou yu laoqu fazhan yanjiu* [A study of Red Tourism and the Old Bases Areas]. Zhongguo caizheng jingji chubanshe.

Jia, H. (2002). Minguo shiqi lvyou yanjiu zhi jinzhan [The development of tourism research in the Republic of China]. *Lvyou xuekan, 4*, 74–77.

Jia, H. (2004). Luelun minguo shiqi lvyou de jindaihua [A note on the modernization of tourism in the Republic of China]. *Shehui kexuejia, 106*(2), 85–87.

Jiji fazhan hongse lvyou [Actively develop Red Tourism]. (2005, February 23). *The People's Daily*, p. 4.

Jin, Z. (2012, November 7). What makes red tourism so popular. *The People's Daily* (overseas edition). Retrieved from http://english.peopledaily.com.cn/90782/8009039 .html.

Jizhe lvxingtuan you xi xingjinguilai [The journalist tour returned from Wuxi with joy]. (1946, May 26). *Shen Bao*, p. 4.

Li, Q. (2010). *Wanshan woguo lvyou touzi guanli de duice fenxi* [The improvement of analysis of tourism investment management strategy] [Unpublished master's thesis]. The Institute of Fiscal Science of Ministry of Finance of the PRC.

Li, X., & Zhao, X. (2001). Bide zhilv yu xinyou zhilu—Kongzi Zhuangzi de lvyou sixiang bijiao [The journey of virtue and the road to the soul: A comparative study on tourism thoughts of Confucius and Zhuangzi]. *Lvyou xuekan, 16*(1), 70–73.

Light, D. (2000). An unwanted past: Contemporary tourism and the heritage of communism in Romania. *International Journal of Heritage Studies, 6*(2), 145–160.

Lin, H. (2011). Lvyouye de fazhan yu jingshen wenming jianshe de guanxi yanjiu [An exploration of the relationship between the development of tourism industry and the construction of spiritual civilization]. *Chanye yu keji luntan, 10*(23), 96–97.

Ling, F. (1994). Qingnian Mao Zedong lvyou jianzhiguan chutan [Inquiry into young Mao Zhedong's view on tourism]. *Dangshi zonglan, 4*, 8–10.

National Red Tourism Work Coordination Group. (2008). *Zhonguo hongse lvyou fazhan baogao 2007* [China Red Tourism development report, 2007]. Zhongguo lvyou chubanshe.

Peng, Y. (2006). *Zhongguo lvyou shi* [China tourism history]. Zhengzhou daxue chubanshe.

Pingshi choubei qingzhu ertongjie [Peking prepares for the celebration for the Children's Day]. (1949, March 28). *The People's Daily*, p. 2.

Pingshi datan kongmiao wuyi huifu youlan [The Temple of Confucius will reopen on May 1st]. (1949, April 30). *The People's Daily*, p. 2.

Qian, M. (2001). *Zhongguo lishi yanjiu fa* [Historiography of China]. Sanlian shudian.

Quanguo guiguo huaqiao lianhehui de jiben renwu [The basic tasks of the Association of Returned Overseas Chinese]. (1956, October 9). *The People's Daily*, p. 4.

Richter, L. K. (1983). Political implications of Chinese tourism policy. *Annals of Tourism Research, 10*(3), 395–413.

Russia draws hordes of Chinese red tourism packages (2015, 13 September). *South China Morning Post*. http://www.scmp.com/news/world/article/1857722/russia-draws-hordes -chinese-red-tourism-packages-retrace-nations

Shambaugh, D. (2002). China's international relations think tanks: Evolving structure and process. *The China Quarterly, 171*, 575–596.

Sofield, T., & Li, S. (2011). Tourism governance and sustainable national development in China: A macro-level synthesis. *Journal of Sustainable Tourism, 19*(4–5), 501–534.

Song xinan fangwentuan [Sending the visitors to the southwest]. (1950, July 2). *The People's Daily*, p. 1.

Sun, B., Liu, C., Zou, G., & Li, K. (2008). *Mao Zedong tan dushu xuexi* [Mao Zedong on reading and study]. Zhongyang wenxian chubanshe.

Tong, R. (2009). Chen Guangfu yu zhongguo diyijia lvxingshe [Chen Guangfu and the first travel agency in China]. *Dangan chunqiu, 8*, 54–56.

Wang, D., & Wang, S. (2001). *Lingdao ganbu lvyou zhishi duben* [Tourism: A reader for cadres]. Qingdao chubanshe.

Wang, S. (2005). *Zhongguo xiandai lvyou shi* [Modern history of China's tourism]. Dongnan daxue chubanshe.

Wang, S., & Zhang, T. (1998). *Zhongguo lvyou shi (shang)* [History of China's tourism: Volume 1]. Lvyou jiaoyu chubanshe.

Wang, S., & Zhang, T. (2003). *Zhongguo lvyou shi (xia)* [History of China's tourism: Volume 2]. Lvyou jiaoyu chubanshe.

Wang, Y. (1944). *Zonghe xuanchuan xue* [Comprehensive propaganda science]. Guomin chubanshe.

Wang, Z. (2008). National humiliation, history education, and the politics of historical memory: Patriotic Education Campaign in China. *International Studies Quarterly, 52*(4), 783–806.

Williams, R. (1998). The analysis of culture. In J. Storey (Ed.), *Cultural theory and popular culture: A reader* (pp. 45–50). The University of Georgia Press.

Woguo shoupi zifei lvxingzhe zai mosike canguan youlan [China's first self-funded tourists visited Moscow]. (1957, January 11). *The People's Daily*, p. 5.

Woguo zifei lvxingzhe dao lieninggele canguan youlan [Chinese self-funded tourists visited Leningrad]. (1957, January 22). *The People's Daily*, p. 5.

Wu, G. (1994). Command communication: The politics of editorial formulation in the *People's Daily. The China Quarterly, 137*, 194–211.

Xia, F. (2004). "Wenhua dageming" zhong de dachuanlian [The Great Rallies in the Cultural Revolution]. *Dangshi zonglan, 7*, 18–24.

Xinjiang zhujun ji zhongnan huabei shaoshu minzu daibiao dijing [Xinjiang military and representatives of ethnic minorities in Central South and North China arrived in Beijing]. (1950, September 29). *The People's Daily*, p. 1.

Yan, F. (1993). *Dachuanlian: Yichang shiwuqianli de zhengzhi lvyou* [The Great Rally: An unprecedented political tourism]. Jingguan jiaoyu chubanshe.

Yang, G. (1994). Huiyi "wenge" shiqi de lvyou waijiao [Memories of tourism diplomacy in the Cultural Revolution]. *Guoji zhengzhi yanjiu, 3*, 50–55.

Yao, H., & Cao, L. (2002). Chen Guangfu kaichuang zhongguo xiandai lvyouye [Chen Guangfu pioneered China's modern tourism]. *Lvyou kexue, 2*, 36–38.

Yi, W. (2009). *Minguo lvyouye huimou: Guolvshe yanjiu* [A review of tourism in the Republican China: A study of China Travel Service]. Yuelu shushe.

Yousheng lvxingtuan juxing diqici huabei lvxing [Yousheng tour group held the seventh trip to North China]. (1936, September 14). *Shen Bao*, p. 10.

Yu, X. (2005). Preface. In Wang, S. (Ed.), *Modern history of China's tourism* (pp. 1–4). Dongnan daxue chubanshe.

Yuan, G. (1998). Lvyou yu jingshen wenming jianshe [Tourism and the construction of spiritual civilization]. *Guilin lvzhuan xuebao, 9*(4), 7–10.

Zhang, B. (1992). *Zhongguo lvyou shi* [China tourism history]. Yunnan renmin chubanshe.

Zhang, X. (2008). Kongzi "bide" lvyouguan yanjiu yu qishi [A study of Confucius' "figural virtue" view on tourism]. *Guilin lvzhuan xuebao, 19*(3), 436–438.

Zhang, X., & Xie, Y. (2005). Deng Xiaoping lvyou sixiang shi woguo lvyou shiye fazhan de weida zhinan [Deng Xiaoping's thought on tourism was the guideline for the development of tourism in China]. *Shehui kexuejia, 112*(2), 120–125.

Zheng, C. (1996). Qing xi laoqu [Deep feelings about the Old Regions]. *Zhongguo fuyun, 10*, 14–15.

Zhong, S. (2013). "Zhonghua remin gongheguo ge minzu tuanjie qilai"—Mao Zedong wei zhongyang fangwentuan tici de qianqianhouhou ["Let All Ethnic Groups of the People's Republic of China Unite:" The ins and outs of Mao Zedong's inscription for the central mission]. *Zong Heng, 10*, 60–61.

Zhou, Z., & Gao, H. (2008). *Hongse lvyou jiben lilun yanjiu* [A study of the general theory of Red Tourism]. Shehui kexue wenxian chubanshe.

Zhu, L. (1945). Zhanhou zhongguo lvyou jihua [The tourism plan in the postwar period]. *Lvxing zazhi, 19*(8).

Zuo, B., Huang, S. S., & Liu, L. (2016). Tourism as an agent of political socialisation. *International Journal of Tourism Research, 18*(2), 176–185. https://doi.org/10.1002/jtr.2044

5 The social space of Red Tourism
The Yan'an case

It was a cold winter night. I was shocked to see Mao Zedong pacing slowly around the entrance gate of the Yangjialing 杨家岭 site. Wearing his blue-gray tunic suit, Mao was meditating with a lit cigarette in his fingers. While I was approaching him, Mao suddenly looked up at me, asking, "Take a photo?" "How much?" I said. "Ten yuan," Mao replied with a very low tone. "How about five? I'd rather take a few photos of you, not with you." Mao gave me a quick nod. Mao was very cooperative, aptly delivering a series of his iconic postures and gestures like in a photo shoot. I handed him a five-yuan bill after photographing. Mao nodded at his suit pocket. Realizing his signal, I put the money in his pocket. All of a sudden, I felt very embarrassed as if I had insulted Mao by giving him a less-than-a-dollar bill for all he had done for this country. Not interested in reading my awkward look, Mao resumed his walk and meditation like nothing happened in the last twilight before the door closed. It was December 26, 2015, Chairman Mao's 122nd Birthday. Given the time difference, it was Christmas morning in the United States. With the difficulty of bringing the Mao impersonator into reality, my research trip of Yan'an in 2015 ended on an absolutely surreal note.

I toured the Revolutionary Museum of Yan'an (hereafter RMY) earlier the same day. A steady stream of visitors was filing through a makeshift memorial in front of the giant Mao sculpture. The crowd consisted of both locals and tourists, but mostly youth. They either were organized by their work units or came on their own accord. The base of Mao's statue was festooned by bouquets, all of which were set there by those visiting, many of whom came with family and children. The atmosphere was solemn, calm, and occasionally interrupted by the sounds of children's giggling and playing.

The two experiences on that day point to such an intricate social space of Red Tourism that a singular theory of propaganda, or of tourism, or even a simple combination of the two seems inadequate to analysis. While Red Tourism is produced not only by the Chinese state, it is also made through spatial articulation as if it were in a theatrical work to the extent that the meaning of the play is produced not solely by the narrative, but more importantly, by *mise-en-scène*, the spatial arrangement of symbols alongside the manipulation of feelings. Red Tourism produces new meanings and imaginations during touristic encounters, which may, in return, impact its own making.

DOI: 10.4324/9781003231783-5

Having said that I would apply Williams' framework of cultural analysis to my account of Red Tourism, I admit that I have so far directed much attention to the ideological articulation and historical reflection of Red Tourism and relatively little to the social dimension. Perhaps this makes previous chapters somewhat dull, but it will also remedy that.

To do so, this chapter focuses on the social space of Red Tourism and examines it through Yan'an. Drawing upon social/critical scholarship on media and space, I argue that Red Tourism is better conceived as a social space, both produced and productive. From the qualitative data derived from my fieldwork spanning a decade, and through a comprehensive spatial analysis of architecture, urban planning, urbanism, and walking tour of the historical site and the RMY, I argue that Red Tourism was created by the state to *xuanchuan*/propagandize its revolutionary past and attached politico-ideological legitimacy by catering to post-socialist nostalgia and stress while producing and proliferating new meanings and imaginations. Departing from Henri Lefebvre's powerful thinking around the production of space, this chapter sheds additional light on the close ties between propaganda and space.

It should be noted that instead of directly applying Williams' idea of a "particular way of life" or "social life" in his cultural analysis framework, I adopt Lefebvre's framework of "social space" for my analysis of Red Tourism instead for several reasons. For one, Red Tourism is a spatial project. Macroscopically, space matters to the production of Red Tourism as development of touristic sites, local economies, and labor markets are all space-oriented. Microscopically, communication between Red Tourism sites and among tourists is largely achieved through a variety of spatial articulations such as exhibits, displays, performances, sculptures, architecture, etc. Secondly, tourists flow across space. Their spatial practice is producing Red Tourism while simultaneously being reproduced by it. I will later elaborate on the term "spatial practice." Suffice it to say that for now spatial practice contributes significantly to the overall dynamics of Red Tourism, which should not be underestimated. Thirdly, there is a connection between Williams' social life and Lefebvre's social space beyond the "social." Both see the relationship between the superstructure and the base not as the latter determining the former but dialectically dynamic, which I have noted in my theoretical exploration of ideology in Chapter 2.

Based on a series of research trips to Yan'an from 2011 to 2021, the current study adopts a qualitative and interpretive approach. The qualitative data are drawn particularly from two strands of inquiry: interviews with government officials ($n = 11$) and with tourists ($n = 32$). The interviews with government officials focus broadly on the development of local Red Tourism, related propaganda campaigns, and urban planning projects. In the interviews at different Yan'an tourist spots, I directly asked "why did you choose Yan'an as your travel destination? Do you like the place you are visiting now? Why? What does this trip/tour mean to you?" The interviews are complemented by other data, including observation notes undertaken in the field trips and a total of 3,222 photos taken by me. Furthermore, I also probed into popular discourse pertaining to Yan'an's Red

Tourism, including articles from national news outlets and social media posts. Last but not least, I analyze government documents and archives – either obtained from public libraries or pulled from the Internet – to gain a fuller picture of the social space of Red Tourism in Yan'an.

In what follows, I start with a critique of what I refer to as the temporal hierarchy of communication study, followed by a literature review centering on the nexus between communication and space. The purpose of the literature review is to introduce the social space approach for the current study by looking into the keywords/key dimensions of said space. Then, I delve into my case study. I introduce Yan'an by focusing on its historical significance and contemporary (re)significations. My spatial analysis of Red Tourism in Yan'an is twofold, containing a macro-level analysis and a micro-level analysis. Macroscopically, I analyze Yan'an's urbanism, urban planning with a particular emphasis on a series of spatial practices pertaining to the representation of space and representational spaces. Microscopically, I examine three representational spaces of Yan'an's Red Tourism, namely, the RMY, the walking tour of Yangjialing, and the *Fantasy* show. Altogether, my spatial analysis aims to illustrate one of the crucial points of this book that propaganda is not a text, a discourse, a technique, but a social space, comprising communication networks, political pathways, and economy. This chapter ends with a discussion of Red tourists' experience characterized by their marked ambivalence towards the grand narrative. This, again, complicates the linear thinking of propaganda and points to a lived, multilayered, and productive social space of Red Tourism.

The temporal hierarchy of communication study

Communication of any kind must attach to a certain space: praying in a church, dancing on the stage, lecturing at an auditorium, chatting on the Internet, casting your mind back to the past. Space – whether physical, abstract, or virtual – has enormous impact on human communication. Massey (1984) has noted that space matters to all social processes. Nevertheless, this space-attached attribute of communication has been largely de-spatialized and systematically overlooked by many strands of scholarship later in the field. A critique given by James Carey (2009) is that "[a] ritual view of communication is directed not toward the extension of messages in space but toward the maintenance of society in time" (p. 15).

The temporal hierarchy of communication study has been pervasive but rarely recognized. I coined the term "temporal hierarchy" to pinpoint a dominant ideology of communication study in which time is the dominant and determining force for communication whereas space is submissive or even dispensable. According to this ideology, acts of communication such as war propaganda and public opinion occur in a sequential order in time, methodologically characterized by a pre–post analysis grounded on a cause–effect model. This was where media effects studies, a dominant paradigm in the field, started. Early scholars in this line of research held different or even contradictory findings and theories in many respects, including time range, strength, and scale of media effects. But they all stuck to the before–after measurement and consequentiality determined by time. Take the development of the hypodermic needle model and the weak/limited effects model for instance. Propaganda matters to scholars only when being considered an event that was solely temporal in an almost unthinkable sense. Under this temporal hierarchy, CPC's propaganda was understood as a legacy of the Soviet Union, both were labeled as Communist propaganda, and finally it was traced back to Nazi propaganda. In this genealogy of propaganda, geography appears utterly irrelevant, despite the fact that the defining words ahead of these propagandas, namely, "Chinese," "the Soviet," and "Nazi" themselves, are explicitly pointing to geographies. Sadly, academic inquiry into propaganda, the great legacy of communication study, suddenly stopped there. It seems to me that the end of the Cold War, again, a watershed and a temporal marker, had rendered propaganda – as a concept, a social practice, or a theory – somehow residual and redundant, and therefore academically worthless. For example, according to this model of thinking, no matter what the research design is and what the approaches are, the outcome of Red Tourism research is highly predictable: once Red Tourism is deemed as propaganda, it, then, can conclude that Red Tourism is a production of the Party's politics being used to spread the dominant ideology. But, among many other puzzles, why has Red Tourism gained so much currency in post-socialist China? The de-contextual, ahistorical, and acultural nature of the ideology of the temporal hierarchy seems to erase all social and cultural texture of Red Tourism, all nuances, all dynamics, leaving only a stubborn "red" stain. This is surely not the direction that I am headed for.

The temporal hierarchy, however prevailing and dominant these days, is not initially how communication as a field was envisaged. Sharing the gene with "commune," "communion," "commons," and "community," communication was once perceived as a spatial project, involving all kinds of networks across all realms of human society. Those networks include national markets and international trading routes in the economic realm, the public sphere in the social realm, and Sacred Congregation for Propagating the Faith of Roman Catholic Church in the spiritual realm, just to name a few. This is to say, communication was believed to be produced by and attached to space. Many scholars have already said this, implicitly if not explicitly. A quick catalogue of the influential authors who have thought about communication from a spatial perspective includes Henri Lefebvre, Benedict Anderson, Armand Mattelart, Umberto Eco, James Carey, David Harvey, Walter Benjamin, Jürgen Habermas, Georg Simmel, Siegfried Kracauer, John Hannigan, and many others. One of the benefits of looking communication as a spatial event from these authors is to locate communication in a larger blueprint of society, which, in turn, helps extend our knowledge about communication not on trivia, but on critical issues/projects such as social networking (Armand Mattelart), nationalism (Benedict Anderson), modernity (David Harvey), lifestyle (Siegfried Kracauer), democracy (Jürgen Habermas), and the economy (John Hannigan).

Space, social space, spatial practice, and spatial analysis

Henri Lefebvre might be a mighty warrior fighting against the temporal hierarchy. Lefebvre (1991) claims that "time was thus inscribed in space, and natural space was merely the lyrical and tragic script of natural time" (p. 95). Then, what is space and what exactly does space mean to communication?

Discussions of space have been throughout the history of philosophy and that of science, not mentioning geography and the great urban–rural division in sociology. Even an extremely brief retelling of any of these can be truly overwhelming. Besides, endless philosophical-epistemological debates on space stray from my topic of Red Tourism. Communication-wise, my inquiry into space focuses on a few terms that show how space communicates. As such, the following discussion is aligned with such authors: Henri Lefebvre, Michel de Certeau, Walter Benjamin, James Carey, and Siegfried Kracauer. My account of the literature on space is intended to be evocative as it serves to provide minimal background theoretical foundation for my later spatial analysis, rather than to be extensive and decisive.

Space is not a place. For de Certeau (1984), the difference between place and space is radical: place, distinctively located, is static and stable, whereas space, consisting of direction, velocities, and time variables, is dynamic and mobile. A place, however, is ready to be transformed into space by spatial practice within that place. For example, a touristic place such as a museum is a place fixed in its geographical coordinates; it turns into a space when tourists visit it. "Visiting" is a spatial practice that distinguishes space from place. Hence, a comprehensive analysis of space necessarily involves two types of examination: one of place and one of spatial practice, or "ways of operating" in that place (de Certeau, 1984, xiv).

The distinction between space and place led de Certeau to identify two types of power in a pair of space-oriented ideas: "strategy" and "tactic." For de Certeau (1984, xix), institutional power rests with a place, which he refers to as "proper," "a spatial or institutional localization." The way to exercise that power was conceptualized as "strategy." In contrast, "tactic" is where the power of resistance arises. For de Certeau, without "proper," tactics are activities within a place. It is such everyday practice as talking, moving, and reading that makes a place a powerful space, resulting in "victories of the 'weak' over the 'strong'" (de Certeau, 1984, xix).

Space is social. This is a core point of Lefebvre's thinking on the production of space. Inasmuch as natural space does have values and carries and delivers symbolic meanings to human beings, the sharp divide between natural space and social space disappears. Lefebvre (1991) adds that "natural space was soon populated by political forces" (p. 48). In other words, all kinds of space are social. The greatest achievement of Marx, as Lefebvre (1991) has noted, is the "successful unmasking of things in order to reveal (social) relationships" (p. 81). Following this tradition, Lefebvre claims that space is not only composed of things, but also social relationships concealed within.

In de Certeau's account, social space is merely a different way to say space as he used the two terms interchangeably. In this regard, Lefebvre's notion of social space was a leap forward. For Lefebvre, social space has specific connotations. Under capitalist conditions, social space connotes "biological reproduction (the family)," "the reproduction of labor power (the working class per se)," and "the reproduction of the social relations of production" (Lefebvre, 1991, p. 32). Lefebvre goes further, suggesting that social space connotes those relations through symbolic representations. And these representations are critical to communication.

At the heart of Lefebvre's spatial representation is a conceptual triad, or "three moments of social space." The first concept is "spatial practice." Lefebvre (1991) points out that spatial practice "secretes" social space (p. 38). "Secrete" herein indicates two processes. First, it implies that spatial practice *articulates* the symbolic meaning of social space in such a *hidden* way as daily routine, or the "practice of everyday life" to borrow a phrase from de Certeau. Lefebvre illustrates this point by an example of the daily life of a tenant. Second, "secrete" means "produce," suggesting that spatial practice *produces* social space. Both processes point to a realm of perception. The second concept is "representation of space." It is "conceptualized" space insofar as it is a system of verbal signs abstracted from the dominant space. The representation of space is framed and ideological in nature, to paraphrase. The third concept is "representational space." It is the dominant space associated with symbolic meanings from which the abovementioned representation of space is derived. In light of dialectical relationships among the three concepts, Lefebvre (1991) refers to the triad as the "perceived-conceived-lived triad" (p. 40).

Among many kinds of spatial practice, travel is a "pure experience of space" (Kracauer, 1995, p.66). Once again, space should not be limited to physical space. It must also include abstract space where individual and collective memory, nostalgia, imagination, ideology, bias, worldview, and the like are constantly being reconfigured, producing the spatial experience of the tourist, and therefore the meaning of tourism. This process is somewhat like reading a newspaper. Different pages and columns (physical space) of the reader's choice take the reader to different topics and discourses (abstract space). This is the reason why Kracauer (1995) contended that "man is really a citizen of two worlds," one he calls "the Here," the other "the Beyond" (p. 68).

To deal with the space of "the Beyond," Benjamin's approach largely centers on imagination. Such imagination is not merely about associating a mental picture with a symbolic value. Rather, for Benjamin (1979), the reading of a city and its architecture involves imagining a dynamic set of human *activities* stirred by an impulse. To a great extent, this imagination is located in the realm of sense, something like what Barthes (1979) referred to as a "dream" (p. 81). According to Benjamin, reading about a city and its people actually means remaking a mental montage by piecing together fragments of memory, feeling, and dream with a unified "leitmotif."

But Lefebvre strongly opposed the treatment of space as a simple mental space. In criticizing Roland Barthes's general semiology of space, Lefebvre (1991) reasoned that the coding and decoding process of the semiological approach reduces space to a message, neglecting history and practice. In other words, "the mental realm comes to envelop the social and physical ones" (Lefebvre, 1991, p. 5). For Lefebvre, spatial analysis, whatever it is, it must emphasize mediation. This is to say, spatial analysis is not only a matter of reading, but also a matter of (re)constructing.

Constructing reality is communication. It manifests in James Carey's (2009) definition of communication, which is "a symbolic process whereby reality is produced, maintained, repaired, and transformed" (p. 23). In that sense, Carey (1989) describes the United States as "the product of literacy, cheap paper, rapid and inexpensive transportation, and the mechanical reproduction of words" (pp. 2–3). Lefebvre would translate Carey's words like this: as a social space, the United States is jointly produced by spatial practice such as "the mechanical reproduction of words," representational spaces such as "cheap paper," and the representation of space such as "literacy" and "inexpensive transportation." By the same token, Carey (2009) sees media of communication (e.g., tourism) as "organisms," which clearly echoes Lefebvre's notion of social space in that both ideas point to a polyvalent nature: produced yet productive.

This is how Lefebvre has paved the way to rethink Marx's political economy of production. For Lefebvre, social space consists of political pathways and communication networks. On the one hand, social space is a product to be used and consumed. On the other hand, it is also a means of production in Marx's term, meaning that a social space is also productive. Lefebvre names this as the polyvalence of social space. This brings up a critical question, then, how are we to analyze social space? A social space is not merely a finished product ready for coding and decoding. It is alive and productive. Neither a semantics of space nor a semiology of space is sufficient.

The answer is hidden in Lefebvre's text. Not only has Lefebvre developed a battery of concepts by which to think through social space, but also a methodological framework that he loosely refers to as spatial analysis. Nevertheless, Lefebvre does not specify the analytical steps of his method. Rather, the sequence of his method can be generated based on his complex and multiplex detail of analyses of urban space throughout his seminal book, *The production of space*.

Lefebvre's spatial analysis takes place in two dimensions to put it roughly. On the micro level, spatial analysis is threefold, including formal analysis, structural analysis, and functional analysis. But these analyses are not enough. Spatial analysis needs to move to another level. On the macro level, Lefebvre analyzes the three moments of social space, namely, spatial practice, representational space, and the representation of space. Spatial practice can be any activities within a given social space, for example, "the practice of everyday life" (de Certeau, 1984) or "a specific use of space" (Lefebvre, 1991). But spatial practice has specific meanings in Lefebvre's analysis of urban space. In a footnote Lefebvre (1991) elucidates that spatial practice in an urban setting takes three levels, namely, "planning," "urbanism," and "architecture" (p. 25). Lefebvre's notion of social space and his spatial analysis serve as my theoretical departure and the methodological foundation for my analysis of Red Tourism in the Yan'an case.

Yan'an: the sacred place

I chose Yan'an as a single case study for its historical significance to the Chinese Revolution, symbolic power for the party-state, and central position in the Red Tourism industry.

Known as the "sacred place of the Chinese Revolution," Yan'an is now a prefecture-level city of Shaanxi Province in northwestern China. As such, Yan'an has heavily invested in Chinese revolutionary heritage. It is a city that is aware of itself as in the center of one of China's poorest regions and yet brands itself as a haven for Red Tourism. For Chinese people, the term "Red Tourism" would immediately invoke a mental montage of a few places, and Yan'an, arguably, may top the list. To the extent that Yan'an represents the turning point of the Chinese Revolution, it is like Gettysburg to the American Civil War and Saratoga to the American Revolution. But Yan'an is much more than that.

Historical significance: the Yan'an period

Yan'an is the site of authentic revolutionary history. In January 1937, Mao Zedong along with the CPC Central Committee were stationed in Yan'an, which became the center of the Chinese Revolution in the following decade until 1947. This decade is known as the "Yan'an period." During this, the CPC drew lessons from the past, advanced its revolutionary theories, formulated the Party's guidelines, principles, and policies, all of which laid the groundwork for the victory of what came to be called the "New Democratic Revolution." Judd (1985) argued that the Yan'an period resulted in the "creative formulation of policies and of methods of organization" (p. 377). What's more, Yan'an represents the maturing of the first generation of Chinese leadership under Mao Zedong and historical transformation of the party from weak to strong (Liu, 2006).

No matter how this history was presented in later stages of socialist China, it would be spurious to reduce it to merely an artifact of Mao's storytelling (e.g., Apter, 1993). In the Yan'an days, Mao tested various ideas on a variety of challenges he faced at the time, with enormous impact on China's political and economic trajectory. First and foremost, Soviet China was established during this period. With the founding of political, financial, and educational systems, what some called "Mao's republic" was established. It is now considered an archetype of the Chinese nation-state. Second, in view of the tentative war against Japan, Mao consolidated his military thinking of "guerrilla warfare" and "mobile warfare," all of which not only became principal strategies for the Communist army in the War against Japan, but also proved to be effective in the later Liberation War against the Nationalists. But Apter (1993) was right in pointing out that "Yan'an was a discourse community" (p. 208). It is so not only because most of Mao's influential writings were produced there, which is true, but also because the most enduring and dominant official discourse of the party developed in Yan'an, noticeably, "criticism and self-criticism," "self-reliance and hard work," "utter devotion to others without any thought of self," and so forth.

Symbolic power: the Yan'an way

The discourse of Yan'an holds enormous symbolic power. I draw the term "symbolic power" from Bourdieu (1989) to designate what he describes as a power of "world-making" (p. 22). In other words, Yan'an has been deemed as a sacred site because of its world-making power in the literal sense. That "world" later referred to as the "New World" of the Communists in contrast to the so called "Old World" of the KMT.

Worldwide, Yan'an signifies the "Yan'an Way." There is no such a term in Chinese, however. The Yan'an Way, to a great extent, might be called the *xuanchuan* way in light of what I have elaborated earlier: that the idea of propaganda is to be imagined as an integration force. Likewise, Selden (1995) defines the Yan'an Way as an "integrated program," stating that:

> [The Yan'an Way] represents a distinctive approach to economic development, social transformation, and people's war. Its characteristic features included popular participation, decentralization, and community power. Underlying this approach was a conception of human nature which held that people could transcend the limitations of class, experience, and ideology to act creatively in building a new China.
>
> (p. 170)

Translating Selden's text would yield the Yan'an Spirit. While definitions of the Yan'an Spirit are varied, its core elements commonly involve self-reliance and hard work, education at the grassroots, and the mass line. These elements well correspond to Selden's "economic development," "social transformation," and "people's war." In fact, Selden (1995) himself, too, loosely refers to the Yan'an Way as a "spirit" (p. 224). In China, Yan'an is believed to connote a starting point of the "way" to socialism with Chinese characteristics (Zhu, 2012).

Two interrelated historical moments named after "Yan'an" are paramount to an understanding of the Yan'an Way or the Yan'an Spirit, and hence, the discourse of Yan'an. They are Yan'an Rectification (*yanan zhegnfeng* 延安整风) and Yan'an Talks (*yanan jianghua* 延安讲话). Credited as the first mass ideological movement of the party, the Yan'an Rectification of 1942–1944 had a profound impact on the CPC and later the PRC. The term "rectification" simultaneously denotes that something was wrong and the necessity of making corrections. In the Yan'an Rectification, the first part of this dual denotation was comprised of a series of isms, mainly, subjectivism, sectarianism, and party formalism, whereas the second part was indeed a method of mass ideological movement characterized by open criticism. In studying the Yan'an Rectification, Western scholarship has been particularly obsessed with immediate effects such as the assertion of the authority of Mao as the leader (Apter, 1993), internecine power struggles within the party (Benton, 1975; Goldman, 1967; Wylie, 1980), a defining moment for Chinese communism (Selden, 1995), and the interconnection between the Rectification and concurrent counterespionage campaigns (Seybolt, 1986). In contrast, Chinese scholars have emphasized the normative power of the Yan'an Rectification insofar as they commonly view the ideological movement as a model (Fan & Hu, 2013; Wang, 2007), a mechanism (Xi, 1999) that can be reused or applied to the Party's building up (Luo, 2012). In fact, the form of the Rectification was duplicated at least in two other political movements in 1950 and 1957. Hence, in the Party's lexicon Rectifications always go in its plural form. In 1996, the CPC embarked on a three-year long ideological education campaign called the "Three Stresses" (*sanjiang* 三讲) that refer to the three imperatives as "stress study," "stress politics," and "stress righteousness." Chinese scholars have generally agreed that this ideological campaign can be understood as an extension of the Yan'an Rectification (Lu, 1999; Luo, 2002; Song, 2001; Zheng, 1999).

Central to the Rectification was the Yan'an Talks, short for "Talks at the Yan'an Forum on Literature and Art"[1] given by Mao Zedong on May 2, 1942. The Yan'an Talks called for the ideological transformation of intellectuals through political reeducation. Rather than all intellectuals, Judd (1985) has noted that the Talks in particular spoke to writers, artists, and dramatists whom Judd refers to as literary intelligentsia. Previously called literati in traditional China, this group of intelligentsias shared considerable propagandistic power for they worked on people's minds and imagination. As earlier noted, this is the defining characteristic of the literati group, for instance, the *shi* group. In a sense, the Yan'an Talks meant to use propaganda to counter other propaganda. In a literary study of He Qifang 何其芳 (1912–1977), a poet, Zhao (2005) described the transformation of He in light of the Yan'an Talks as "from being 'enlightened' to 'enlightening' the masses" (p.10). In that sense, the Talks can be better understood as the groundwork for the party propaganda work than what Irving (2016) viewed as the foundation of China's policy on culture. In other words, while the subjects of the Talks were ostensible literature and art, they were really an exercise in formulating ideology, thusly propaganda in a deeper sense. A marker of the Yan'an Period, Lu Xun Academy of Arts is a perfect footnote in this regard; the Academy was built not because the CPC had a keen interest in arts in the middle of the war but rather because it helped prepare a new generation of propagandists.

The Yan'an Spirit continues to be studied and celebrated in China today. Yan'an is seen as the ultimate source of positive energy for the Chinese state. There is even an area of study called "Yan'an Studies," which was established in the early 1990s in China (Guo, 2006). With regard to Red Tourism, the latest metaphor of Yan'an is a "test area" for later People's Republic. Compared to the early signifier – either the "Yan'an Spirit" in China or the "Yan'an Way" world-wide – the new metaphor of "test area" is spatially illuminating and imaginatively appealing; it literally invites tourists to the "area."

The "Red Mecca": a paradoxical utopia

What some call the "Red Mecca" was and is a utopia to many, and it is a paradoxical one at that. The paradox manifests in a series of sharp contrasts between abstract space and physical space throughout the recent nearly a hundred-year history of Yan'an. During the Yan'an Period, thousands of intellectuals from all over the country came to the city in pursuit of the revolutionary ideal despite its geographical status as one of the poorest regions in the nation. In the 1960s, the youth flocked to Yan'an in the name of revolution at a time when many party leaders of the Yan'an Period were being persecuted in the Cultural Revolution. Finally, when the so-called "digital revolution" became a reality and Chinese netizens began indulging themselves with activities in virtual space, when capitalism as a mode of production and exploitation became dominant in economic space, and particularly when Hollywood-like cultural products and outbound tourism increasingly gained popularity, Yan'an is now surprisingly re-emerging as one of the nation's most popular touristic destinations.

The Red Tourism industry without Yan'an is unimaginable, and vice versa. Branded as the "most condensed, most resourceful Red Tourism city" on the government's website, Yan'an is the host of 445 revolutionary sites in addition to three big ideological education bases for patriotism, revolutionary traditions, and the Yan'an Spirit (Yan'an Municipal Government (YMG), 2013). According to a government report, in 2015, Yan'an's tourism sites received 35 million visitors, with an annual growth of 11.3 percent; the consolidated revenue of tourism was RMB 19.3 billion ($2.97 billion) with an annual growth rate of 12.1 percent (Statistical Bureau of Yan'an City, 2016). The same year, the gross output value of Yan'an's primary industries (including agriculture, forestry, animal husbandry, and fisheries) was RMB 19.8 billion ($3 billion) with RMB 15.4 billion ($2.37 billion) in agriculture. Given that traditionally Yan'an was built up on an agricultural-based economy, it appears now that the contribution of Yan'an's tourism to the local economy has outstripped its agriculture. In the last 15 years from 2005 to 2020, Red Tourism revenues and tourists in Yan'an have shown a steady positive growth, except a dramatic decrease in 2020 due to the COVID-19 pandemic (Figure 5.1).

Spatial analysis I: on the macro level

Yan'an once represented a living hell. Ritualistically, this is how various accounts of Yan'an start. Yan'an was described as the place that "comes the nearest to being a liability" in Mark Selden's (1995) book, *China in revolution*. Similarly, in Chinese narratives, the image of Yan'an before the Yan'an Period has been exclusively associated with barren lands, warlordism, banditry, and famines. Perhaps, Yan'an is best known in the West as the Red Capital in American journalist Edgar Snow's 1938 book *Red star over China*. Even in this book, Snow (1938) rarely mentioned Yan'an. Only in a footnote, Snow told readers that Yan'an "is now (1937) the provisional Red capital" (p. 28). It was called "the

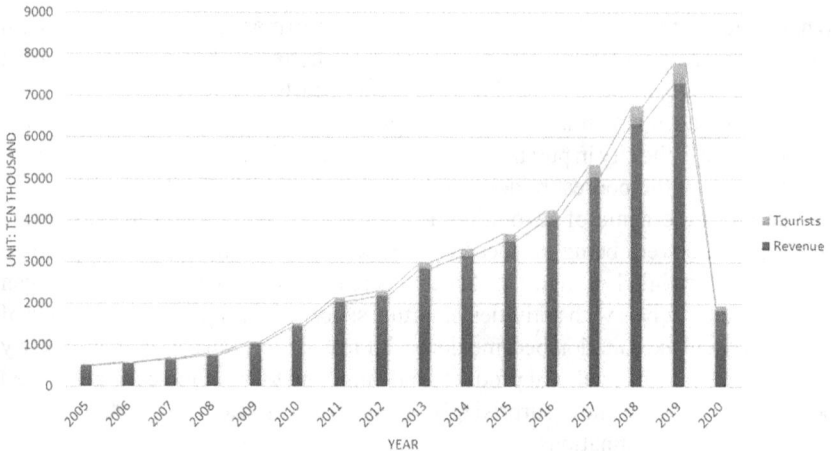

Figure 5.1 Red Tourism revenues and tourists of Yan'an from 2005 to 2020. Created by the author. Source of data: Statistical Bureau of Yan'an City, http://www .yanan.gov.cn/

city made of earth" in a report of a major newspaper in the KMT period (Yan'an jian ying, 1947). This was also my impression before I visited in 2011. But I was wrong.

Urbanism: a hybrid space

I was stuck at the busiest road outside of the Yan'an Railway Station in August 2016. This happened again and again during my multiple trips to Yan'an in the last few years without exception. It is almost impossible for a tourist to get a taxi in the vicinity of the station. This is, however, common in China's "first-tier cities" such as Beijing, Shanghai, and Shenzhen. However, they are defined, these cities are characterized by high-density housing and terrible traffic congestion. Yan'an shares these too. Except Yan'an is not a first-tier city, not even a well-qualified third-tier city by criteria such as population, income, economic competitiveness, etc. For example, Yan'an had a total population of 2.23 million by 2015 (Statistical Bureau of Yan'an City, 2016), which is the equivalent of a county in Anhui province.

When I finally got into a taxi, the driver took detours to avoid traffic. From my accent and backpack, the driver could tell that I was a tourist. Nationwide, thousands of taxi drivers in several cities had staged protests and strikes against unfair treatment and ride-hailing app services since the beginning of the year. Ironically, due to frequent poor service from cab drivers, Chinese netizens had little sympathy for cab drivers, quipping, urging them to "Please don't stop striking!" on social media.

The long, unexpected scenic journey made by the taxi driver, however, was like a tour of the city to me. It was packed with a montage of a global scene, a modern scene, a local scene, a metropolitan scene, and a rural scene. Pizza Hut and KFC restaurants occupied the most eye-catching locations in the city's business area hand-in-hand with the newly erected Baskin-Robbins shop. Note that these global fast-food restaurants and ice cream chains normally open their stores only in first-tier or second-tier cities in China following their middle-class clientele alongside their presumed tastes. Rarely would they pick a small city the size of Yan'an. This exception is no doubt due to the city's Red Tourism boom. From my observation, those who frequent Western chain establishments were mostly tourists. The locals preferred to hang out in those Chinese restaurant chains not so much because of their marked preference of rice and noodles but because of something called *xing jia bi* 性价比, a buzzword referring to a kind of cost-effectiveness related to everyday consumption. Also note that a regular 12-inch medium pizza is normally sold for **88** yuan ($13.5) in Chinese Pizza Hut stores, more expensive than in the United States at the time. Presumably, the preference of *xing jia bi* can be an indicator for lower-income groups in China. Outside those modern-look shopping centers and malls, local villagers were peddling yellow steamed buns. The previously unknown specialty of Yan'an has caught national attention overnight as it was featured in an episode of *A bite of China*, a smash-hit China Central Television (CCTV) documentary series on Chinese local cuisine. More villagers gathered in the major tourist sites of the city such as the Date Garden and Yangjialing, selling dried dates, another local treat of Yan'an, to tourists. A report of the *People's Daily* published a decade ago criticized the peddling around these Yan'an historical sites as "inharmonious voice" (Wang & Wang, 2005). A date market was built right next to the entrance gate of the Yangjialing site but villagers continued peddling around. But this was no longer the case when I revisited the city in 2021. Peddlers are all gone, a leap forward in modern capitalized tourism.

The hybridity of Yan'an is hard to decipher and represent. Yan'an is not rural. Fast-paced urbanization has been the dominating theme of the city. During the Eleventh Five-Year Plan period (2006–2010), the built-up area of central Yan'an city doubled (Zhang, 2011). A few years ago, the sign of the handwritten Chinese character *chai* 拆 (demolish) inside a big circle appeared in many walls of old buildings in the central parts of the city. The sign articulates a huge ongoing modernization project that Yan'an shares with many other small cities in China. But modernity does not simply mean destruction of the old; it has to be "creative destruction," a radical break in lifestyle, according to David Harvey (2003). In other words, modernity, to a great extent, is more about urbanism than urbanization. Yan'an is not global either. Hardly can anyone spot a foreign tourist in Yan'an despite its status as a popular tourist city. According to the government statistics, international tourists only accounted for 0.1 percent of the total tourists that Yan'an received in 2015 (Statistical Bureau of Yan'an City, 2016). This is why there is no international hotel chain in Yan'an. In the absence of chained department stores, local ones are glutted with cheap products.

The representation of space and the representational spaces

The representation of space and representational spaces of Yan'an can be simplified as what Li et al. (2011) phrased as "two sacred [sites] and two yellow [scenes]." The first part connotes the representation of space as the sacred site for the Chinese Revolution and the sacred birthplace of Chinese civilization. The second part implies two representational spaces, the Yellow River and the (yellow) Loess Plateau.

Located in the middle reaches of the Yellow River on the Loess Plateau, Yan'an along with its periphery is considered the cradle of early Chinese civilization. Once a small town, Yan'an's representational spaces were its yellowish soil (the Loess Plateau), the yellowish water of the river (the Yanhe River), and its yellowish people (typical Han ethnic Chinese). In concert with the representational spaces, the representations of space of Yan'an were hard work and plain living. Taken all together, traditionally Yan'an as a social space was highly expressive and communicative, signifying the greatness of the Han Chinese and their civilization.

This spatial connotation of Yan'an was (re)shaped by the Chinese Revolution. The replacement of "sacredness" for "greatness" reflects the new codification of the representation of space and the representational spaces. The representational spaces became Pagoda Hill (*baota shan* 宝塔山), associated with Mao Zedong's multifarious activities, anecdotes, and iconic photos during the Chinese Revolution, and *yao dong* 窑洞, a unique cave-like indigenous dwelling that accommodated the Communist troops during the revolutionary period. This resignification of Yan'an became a blessing for the city's skyrocketing economy of Red Tourism. Pilgrim-like tourists, young and old from every corner of the country, swarm into Yan'an desperate to have a picture with the Pagoda Hill, to stay a night at a *yaodong*-style hotel, and, most importantly, to get a flavor of the Chinese Revolution. David is an entrepreneur from northern China. Yan'an has special meaning to him. He says:

> I love Yan'an. I was brought up under the Party's red flag, always feeling attached to Yan'an, the sacred place. This is my first visit. Lao Mao [a nickname for Mao] always has a special place in my heart and no one can replace him ... [During the Chinese Revolution] many aspiring young people from all over the country came here to join the revolution. If I were born in that era, I probably would also do the same. I will talk about my Yan'an experience with my friends for sure.
>
> (interview, June 23, 2013)

This is how most of the interviewees spoke about their motivation for visiting Yan'an, depicting themselves as some kind of pilgrim. David was not alone on his pilgrimage to Yan'an. Red Tourism captivates millions of the "Red" pilgrims all over the country each year.

Urbanization, gentrification, and extreme makeover

Like many holy sites, the locality of the Sacred Place is unique. The layout of the city of Yan'an is described as a big "Y" or "three mountains sandwiching two rivers,"[2] which indicates a great scarcity of land (see Figure 5.2). It is also the reason why the spatial development of the city has been described by "lines" rather than by areas. A significant number of locals still live in earthen caves, partly because of the skyrocketing cost of housing and partly because of persistence of the traditional way of life. With the dramatic, ongoing advance of urbanization, the close-knit *yaodong* communities were being disintegrated, rapidly becoming rubble and debris (Figure 5.3). I visited several such sites in last few years, and they resembled ghost towns in the mountains, giving a view of the sharp contrast between untouched well-paved roads, stone steps, and the ghastly debris.

The ghostly scene was a byproduct of an ongoing makeover of the city. In 2011, Yan'an was at the midpoint of the intermediate stage of urbanization, characterized by myriad growing pains such as rapid population growth, land scarcity, traffic congestion, poor infrastructure, tacky urban scaping, and so forth (Qing et al., 2014). To resolve these aesthetic issues, Yan'an underwent a huge Red Tourism-oriented beautification project, which was given the name of "Beautifying Yan'an."

A five-year plan from 2011 to 2015, the project aimed to rebuild a sacred Yan'an exactly like the one imagined by the public. In other words, it was about to create an image of the city that was based on and caters to people's perceptions of the city. The criteria were a triad: beautiful functions, beautiful images, and beautiful contents. The outcome of the makeover was envisioned as that of a "cultural image" which enabled the tourists' pursuits of "bringing back Red memories" and "getting the feel of the Loess style" (Housing and Urban-Rural Development Bureau of Yan'an, 2011). According to the urban planning authority, the end of

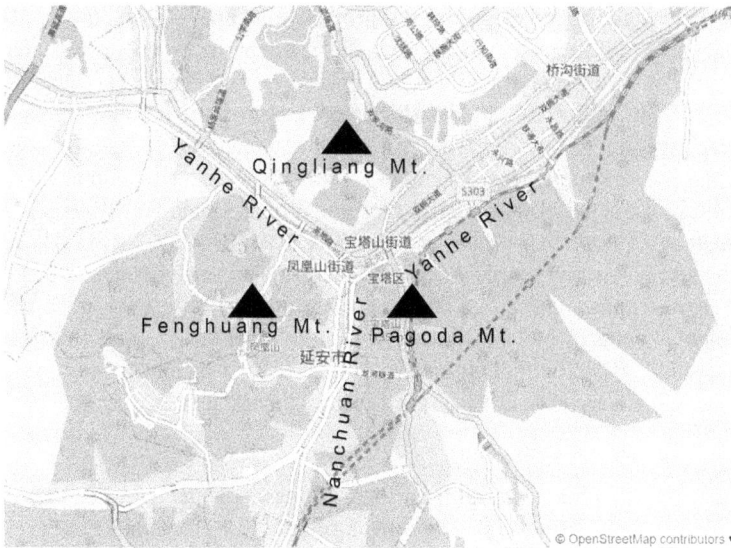

Figure 5.2 Schematic diagram of Yan'an city. Created by the author. Source of data: OpenStreetMap, https://www.openstreetmap.org/copyright

Figure 5.3 An evacuated community. Photograph by the author.

the effort is to significantly improve the taste of the city of Yan'an (Yan'an Urban Planning and Design Institute (YUPDI), 2011).

Whatever it is, the new taste was believed to stem from a Red Tourism-centered gentrification project. The major part of the project was to pave an estimated 30 hills, so that the government could relocate the residents from the city center to the outskirts to make space for its Red Tourism development. The slogans for Yan'an urban planning are "Evacuating People from the Center to Expand the Periphery" (*zhongshu waikuo* 中疏外扩) and "Rebuilding the City on the Mountains" (*shangshan jiancheng* 上山建城). Note that the paved mountains are not in the center of the city. In the central area, the project called for "Relocating Residents from the Mountains" (*jumin xiashan* 居民下山) (YUPDI, 2011). These slogans appeared in various sources including government documents and news reports. To free up space on Pagoda Hill, the landmark of Yan'an, for example, the government evacuated 148 families, totaling 1,100 former residents (Yan'an Tourism Bureau (YTB), 2012). This ambitious makeover was meant to sharpen a sense of sacredness by adding vast space to the historical sites. Sara, a local official, explains:

> Adding space is a philosophy of traditional Chinese architecture. For example, Buddhists would like to dislocate their temple from the residential area of the city to a remote place with large space to make the temple look otherworldly. This is the best way to rebuild the image of Yan'an's Red Tourism.
>
> (interview, June 25, 2013)

Significantly, Sara's understanding of the reconfiguration of space implicitly suggested that there was a similar strategy being deployed between religious practices and Red Tourism in representing a sacred image.

The relocation was an extremely critical element of the city's Red Tourism ambitions. In 2011, the government set out a spatial development strategy for Yan'an city: "Developing the New City, Easing the Old City, and Protecting the Sacred Place" (Qing et al., 2014). To translate the slogan, it simply means a resettlement plan for residents of the old city to relocate elsewhere so that their former homes can be remade to tourist sites. In a research report on the beautifying project, Zhang (2011) argued that the "absolute authority" of Red sites must be asserted and established in urban planning. To do so, as Zhang suggested, it had to put Pagoda Hill at the center of the spatial structure of the city with other Red tourist spots radiating from it. This would achieve a desired effect in urban spatial structure that is termed as *zhong xing gong yue* 众星拱月 in Chinese traditional architecture, literally, "many stars circle a moon" (Zhang, 2011, p. 4).

Beautification was only the first step on the roadmap. In 2012, the Master Plan of Yan'an City, 2012–2030 was introduced. The primary aim was to build an eco-friendly, livable, vibrant, sacred Yan'an by making the city beautiful, big, and strong (YMG, 2012). The strategy is a two-part plan, first sanctifying the old city, then modernizing the new city. The way to sanctify, as earlier noted, effectively was to remove residents from the center of the city to save space for Red Tourism sites. Hence, land being cleared in this process will be only used for Red Tourism and related public green space. This plan, on the one hand, was expected to lead to the restoration of the historical scene of Yan'an characterized by lower buildings, large tracts of open land, and low population density. On the other hand, as the residents continue moving into the outskirts of the city, Yan'an will expand to two-and-half times its current size by 2030.

Many of these urban planning projects vis-à-vis Red Tourism came to fruition during the last decade. Yan'an had noticeable aesthetic improvements on my visit in June 2021 in terms of cleanliness, available public green space, gardens, and walking trails along the bank of the Yanhe River. There are more options for dining, lodging, and entertainment for tourists. Top hotel chains, high-end supermarkets, chic bars, and elegant dinners appear in the business center of the city like any other top tier metros (Figure 5.4). As for gentrification, the moving-to-the-mountain plan seems fairly fruitful at least from the perspective of a real estate developer. For example, the price for properties in newly developed areas doubled, from 6000 yuan/square meter to 12,000, becoming the most valuable of the city and is known for being the homes of the well-to-do. All of these changes have been largely brought about from the city's Red Tourism business. This is to suggest that previously a project, now the "Red industry" is a perceivable reality.

Spatial analysis II: on the micro level

At the heart of Red Tourism is imagination. Benedict Anderson's book *Imagined communities* is helpful on this topic, particularly in thinking about how Red

Figure 5.4 Bargoers walking outside a chic bar in Yan'an. Photograph by the author.

Tourism strives for rebuilding and branding an intended image of the socialist country via reconfiguration of space. For Anderson (2006), the public perception of the nation is largely based on the reconfiguration of space via imagination. This is also true for the Yan'an case. Tourists imagine themselves as part of the nation through a series of institutional apparatuses associated with Red Tourism. The museum, in particular, is of vital importance among those apparatuses.

The Revolutionary Museum of Yan'an: articulation of the grand narrative

Museums are profoundly political. Museums in any form, by any theme, and under whatever situation are not innocent and are by no means neutral and purely scientific. The museum is a space fraught with political/ideological signification. It articulates and spreads the intended ideology through a series of even more complex and subtle apparatuses, including spatial organization, ordering of exhibits, proscriptions, internal and external layouts, decorations, inscriptions, selected artifacts, displays, lighting, and so forth. Like Anderson, Tony Bennett (1995) also viewed the museum as an institutional apparatus, noting the co-occurrence of the reorganization of the social space of the museum and the emergence of the bourgeois public sphere in the West. Bennett (1995) argued that the museum has gradually and ultimately transformed into a governmental instrument for educating the general public ever since its Enlightenment conception. In studying the "Newseum," Nimkoff (2008) demonstrated that this news museum acts as a "mechanism for changing individual minds" (p. 30). Nevertheless, this educational function of the museum has a specific meaning in Red Tourism: ideological education. The RMY is a propaganda apparatus in this regard. To a great extent, the RMY is exactly like the Party's organ. Leon and Rosenzweig (1989) remarked that sponsorship and financing have significant effects on what museums say. A reflection on Herman and Chomsky's propaganda model suggests the same controlling mechanism as in news media.

The post-1949 reconsideration of museums as education resources in China entailed a significant transformation of earlier cultural strategies. However, the rules and admission fees of China's museums in their early stages of development did not fulfill the propagandistic role because they served, though involuntarily, to distinguish the intelligentsia by excluding the general populace. Coincident with China gaining global economic power in the early 21st century, the Party's re-realized the propaganda function of the museum, which led to the free admission and renovation movement for all public museums. During this period of time, the RMY, originally erected in 1950, was relocated to a new site, reconstructed with total expenditure amounting to approximately 570 million yuan ($95 million), and reopened to the public with free admission in 2009. The new RMY was branded as the "Number One Project" (YTB, 2012).

The RMY sits in the Pagoda District, the central tourist area of the city. Through exhibits and extensive collections, the 14,500 square-meter (156077 square-foot) museum presents a 13-year-history of the CPC in Yan'an. In addition to 5,500 historical photos, its collection features more than 25,000 artifacts, from daily use items of the party's early members to Mao Zedong's pistol and his white horse, long dead and now stuffed. The museum includes four connected exhibition units with each showcasing a particular theme (see Table 5.1). Put together, the RMY produces a grand narrative of the party. Note that the RMY closed in January 2021 for reworking of exhibitions and remains closed as of the completion of this book. Two small exhibitions about the party's anti-corruption history in Yan'an and the Long March, however, are still open to the public.

Instead of examining the RMY exhibit by exhibit, I only highlight some nodal points pertaining to the nexus between propaganda and space. A truly comprehensive analysis of the RMY is a book in of itself. Besides, the museum is not the primary focus. In what follows, I start with architecture as it is the obvious center of spatial analysis. Eco (1997) regards architecture as a mass communication system. For him, like any other form of discourse, architectural discourse starts with commonly accepted rationales that accordingly lead to readily acceptable argument. Succinctly, architecture is where the grand narrative begins.

Corresponding to the grand narrative is a grand display. Situated in a large public area, the façade of the RMY is both traditional and modern. On each flank of the main entrance, there are nine statues of iconic Chinese literary characters in their revolutionary moments. This arrangement of the numbers connotes a principle

Table 5.1 The exhibition plan of the RMY. Created by the author.

Unit 1: Destination of the Long March
    1.1 Northwestern Revolutionary Base
    1.2 Headquarters of the Chinese Revolution
    1.3 Formation of the National United Front
    1.4 Political Center of the Anti-Japanese War
    1.5 Comprehensive War of Resistance
    1.6 Anti-Japanese Democratic Bases behind Enemy Lines
    1.7 Policy of "Resistance, Unity, and Progress"
    1.8 Victory of the Anti-Japanese War
Unit 2: Test Area of the New Democratic Revolution
    2.1 Construction of Political Power
    2.2 Economic Development
    2.3 Cultural Development
    2.4 Building the Cadre System
    2.5 Foreign Affairs in Yan'an
Unit 3: Birthplace of the Yan'an Spirit
    3.1 Anti-Japanese University Spirit
    3.2 Rectification Spirit
    3.3 Zhang Side Spirit
    3.4 Bethune Spirit
    3.5 Nanniwan Spirit
    3.6 Spirit of the Comrades of Yan'an County
    3.7 Model Worker Spirit
Unit 4: Guiding Role of Mao Zedong Thought
    4.1 Formation and Development of Mao Zedong Thought
    4.2 Yan'an Rectification Movement
    4.3 7th National Congress of the CPC
    4.4 Struggles for Peace and Democracy
    4.5 Commanding the National Liberation War

of *feng shui* 风水: *jiu jiu gui yi* 九九归一, literally, "every time number nine appears, the next number will return to number one." The figurative meaning is that everything in nature is in a circle and they have to return to their original point ("one") after reaching their climax ("nine"). The key point in this architectural metaphor is the "one," designating the original point of the Chinese Revolution. It is the Party that these 18 sculptures point toward and that the people represented in these sculptures champion. Prima facie, the central sculpture in front of the main entrance is the "one," which, unsurprisingly, is Mao Zedong, symbolizing "oneness" of this whole spatialized *feng shui mise-en-scène* (Figure 5.5). From another angle, however, the RMY façade is embracing modernity: the lighting of the RMY alongside the European-style ornamental horticulture of manicured trees and lawns, arched walls, and stylized layouts appear self-evident. This combination of incongruous architectural motifs from traditional Chinese culture and the West mirrors the contradiction of architecture as a mass communication apparatus to communicate or, more precisely, to advance an argument crystalized in the party's long-run slogan: "Without the CPC, There Would Be No New China."

Figure 5.5 The façade of the RMY. Photograph by the author.

The architectural codification goes on as the grand narrative unfolds. The steps to the entrance are divided into three stages that represent the three historical stages of what the Party experienced in the Yan'an period, namely, the Agrarian Revolution, the Anti-Japanese War, and the Liberation War. The seven *yaodong*-shaped archways in the entrance porch denote the Seventh People's Congress of the CPC held in Yan'an. The interior colors of the exhibition units were initially coded to denote different themes. According to Wei Chunxue (1999), the chief designer of the RMY, grayish purple was used to symbolize the dawn of the revolution, light coral red for the nation's life-and-death struggle against Japan, olive green for the Communist-controlled peaceful border region, light caramel for the Large-scale Production Movement and the Ratification Movement, and Chinese red for the victory of the Chinese Revolution and the founding of the PRC (Wei, 1999). However, when I started this research project in 2011, the interior color of the RMY was all painted Chinese red.

Upon entering the RMY, visitors are greeted by a group of statues of the early Party's leaders along with foreign Communist sympathizers against a giant relief. The open and separate entrance hall where the statues stand is a first-of-its-kind for the Chinese museum. The representational spaces of Yan'an, including Pagoda Hill, the Yanhe River, the Loess Plateau, and the old Yan'an city, were carved in relief to represent the Yan'an period. To hint to the visitor to switch mental space from the present to the past, a special lighting effect – *zaojing* 藻井 or caisson ceiling – was borrowed from ancient Chinese architecture for the relief.

Typically found in sacred places such as temples, *zaojing* is a skylight usually above a religious figure. Remolding the *zaojing* structure, the architect set 13 groups of lights above the relief on the celling. In doing so, it "transforms an unknown space into one that is full of energy and solemnness" (Wei, 2000, p. 53). The spatial organization of displays also glorifies the Party's past and its leadership. In representing the victory of the Anti-Japanese War at the end of the first exhibition unit, for example, three rifles with haversacks across the weapons as the symbol of the Party's revolution are placed in a glass display case in the center of the room with every other image and object enclosing it.

The new RMY promised to depart from the insipid and banal image of Chinese revolutionary museums. Its planners have fulfilled the promise in some respects. In representing the people's War of Resistance against Japan, 20 Kung Fu spears are arranged against the wall in a fan-pattern like rays emanating from the sun. The spears look like rays and the negative space around handles of the spears forms a half-sun-shape. Six rifles were added in the radial pattern. While the display is rich in symbolism, its visualization is so spectacular that it looks like an art exhibit. Normally, visitors crowd into the sight, taking plenty of pictures.

In the course of the chronological progression of the second exhibit on the New Democratic Revolution, foreign affairs in Yan'an are spotlighted. A golden, hollow globe is placed at the center of the exhibition room. Names of those countries with people coming to Yan'an during the Yan'an period are shown and marked red, leaving the rest countries' names blank. The surrounding walls are hung with historical photos divided in three sections, namely, "International Friends in Yan'an," "The US Military Observer's Mission in Yan'an," and "From Yan'an to the World." On a typical day, this exhibition room swarms with giddy kids curiously hanging around the golden globe. It is worth noting that like in any other museum, children along with their family are a staple presence at the RMY.

The exhibit of the New Marketplace of Yan'an received tourists' praise. Established in 1939, the New Marketplace soon became the center of commerce and finance of Yan'an at the time, particularly after the Bank of the Border Region and the Finance Department moved in. Now a three-dimensional simulacrum depicting a life-size barbershop, a blacksmith's shop, among others, filled with wax shopkeepers and customers, the Marketplace occupies a large fraction of the second floor of the museum. The vaulted ceiling is painted with a realistic blue sky and vivid white clouds. The realism of the indoors outdoor space is reminiscent of the iconic Grand Canal at Las Vegas, itself an indoor replica of the Vancian canals. The entire Marketplace is bathed in soft and dim lighting, leaving tourists to fill the unseen details of past, and practically, to enable them to see the faint light inside the shops and draw attention to the displays. To some degree, the Marketplace serves as a "buffer zone" where tourists can relax and contemplate after engaging the comparatively intense discourse on the War of Resistance and before moving on to the next exhibit, the most important one about the Yan'an Spirit from the planner's point of view. Walking through the street of the Marketplace, tourists see a prosperous Yan'an brought about by the Communist force.

Annie, a college student from Guangzhou, appreciates the intriguing style of the RMY when she says:

> It is well worth watching. My only regret is that I would have arranged more time for this museum. It has so many floors. I did not expect that a small town like Yan'an could build a spectacular modern museum like this one … I was most impressed by the high-tech based replica of old shops and streets of Yan'an inside the museum, very realistic, particularly with cutting-edge sound and light technologies.
>
> (interview, June 24, 2013)

What struck Annie most was how the RMY communicates to her as a space: its magnificent appearance and the state-of-the-art exhibit make the revolution-themed museum compelling to the senses.

While the four historical exhibition units are unremarkably arranged, two exhibits stand out to visitors thanks to their noteworthy spatial articulations. Both appear in what the designer calls "remaining" space, a 90-degree corner that links two exhibition rooms. As open space, the corner is meant to "develop rhythms of space to reduce the visual fatigue of the audience," and as transition space, it is a continuation of the last exhibit and a foreshadowing of the next (Wei, 1999, p. 68). But the effect is, perhaps, more profound than that. Leaving the third unit of the Yan'an Spirit before entering the fourth on "Mao Zedong Thought," visitors walk into a corner where a lightbox wall appears. It was supposed to put the state leaders' comments on the Yan'an Spirit on display. Instead of highlighting the words, however, the light boxes featured portraits of the PRC's state leaders. In 2013 summer, there were four portraits on the lightbox wall, namely, Mao Zedong, Deng Xiaoping, Jiang Zemin, and Hu Jintao. Compared to the standard speech photos of the other three leaders, Mao's picture, presumably taken after 1949, was exceptionally refreshing. What caught the visitor's eye is a typical image of a traditional Chinese intellectual writing diligently on paper with a brush. In the background of the picture, there is a wooden window with a green curtain pulled back and sunlight streaming in. As the visitor's eye moves through the window, a big green tree is bathed in glorious sunshine. But the lightbox display had changed when I revisited the RMY in 2015. With Xi Jinping's photo being added to the portraits, the display has now become a representation of the five generations of Chinese leadership. In addition, Mao's picture was replaced by a similar speech photo in concert with others, resulting in a visual homogeneity in terms of frame, posture, foreground and background, etc. Also, the lights above the lightbox wall were turned off, so that the tourist can see only the five portraits.

The other surprising exhibit is at the corner right before tourists move on to "Victory of the War," the last exhibit of the museum. It is a new panoramic photo of Yan'an city, representing a modern Yan'an. Typically, tour guides would stop there to promote the city to tourists. But unlike other exhibitions in the RMY, this one is rather a commercial maneuver as manifested in the tour guide's promotional tone and the PR effort of the city by putting a modern panorama in a history museum. All the surprising elements and changes suggest that the RMY has been constantly adjusting its exhibits, not so much to represent a revolutionary past as to negotiate, reinterpret, and most importantly, to mobilize that history for *xuanchuan*.

The physical space of the RMY is highly controlled. The visiting route is unidirectional to reinforce the RMY's guiding role in constructing the grand narrative of the Chinese Revolution. Visitors' flow through the museum is on a predetermined route without allowing for any variance. In other words, tourists do not have navigational autonomy as in other kinds of museums, and must proceed along the approved path. In controlling visitors' flow, the planners hope to construct a kind of filmic reality in which desired interpretations are cued by pushing the visitor from one particular exhibit/display to another, as if in a film sequence. In order to enhance the impact of the exhibit, the planners have also compressed the exhibit space by reducing the floor level from 6 meters to 4.8 meters and lowering the exhibition wall from 4 meters to 3.5 meters (Wei, 1999).

While taking pictures is strictly prohibited in many history museums in China for conservation reasons, photography and video recording are surprisingly welcomed in the RMY. Through the use of mobile devices and social media, the tourists, previously considered recipients of propaganda messages, now act as propagandists when they share intended images and text to much wider audiences. This propagandee-*cum*-propagandist role of the consumer manifesting in the changing process between consumption and production can be better understood by the concept of "media conduction," because as Peaslee (2013) pointed out, "it sees that relationship as defined by processes"[3] (p. 824). To put it metaphorically, by applying the original notion of conduction, the audience now becomes a "conductor" through which "electricity," or ideology-loaded messages, can pass along.

During the museum tour, tour guides always highlight two anecdotes. The first is usually referred to as "Voting with Beans." It illustrates the democratic structure of the Shaanxi–Gansu–Ningxia Border Region Government, the most embryonic form of later People's Republic. The on-site narrator stops in front of a historical photo in which a group of people cast beans in a row of bowls on the table. The narrator explains the scene:

> The highest authority in the Shaanxi-Gansu-Ningxia Border Region was the parliament. According to the election law, members of parliament were chosen by the masses through the parliamentary election. Because the literacy rate at the time was only about one percent with most being illiterate, they put a bowl behind each candidate, so that people could vote by casting their beans in the bowls in public. This method was democratic and transparent.
>
> (July 13, 2014)

Note that terms like "parliament" and "parliamentary election" are so foreign to the Chinese audience that they can immediately invoke an imaginary of Western democracy. A typical comment from visitors is, "How democratic it was back then!"

The second is a murder case, called the "Huang Kegong Case." It was the first murder case of the High Court in the Border Region. An exceptional young general of the Red Army during the Long March, Huang Kegong shot Liu Qian, a female student of the Anti-Japanese Military and Political University in Yan'an after she rejected Huang's marriage proposal. Huang was sentenced to death after a public trial, the first influential one in the Party's history. Mao Zedong refused to grant an amnesty for Huang, saying "members of the Communist Party and the Red Army must maintain more stringent disciplines comparing to the ordinary people." This story was adapted into a film in 2014. With an English title *A murder besides the Yanhe River* (Chinese title still *"The Huang Kegong case"*) and starring Wang Kai, a *Marie Claire* cover boy and a valuable television series drama star, the film was touted as a "political thriller" by *China Daily* and a "rare 'Red' commercial film" by its director (Wang, 2014).

The walking tour of Yangjialing: searching for the lost Xanadu

Unlike analysis of the RMY, my examination of the Red Tourism sites of Yan'an does not focus on displays, artifacts, and images. Instead, I shift focus to spatial practice. Tourism is, for one, about touring by and large, and that is a spatial practice. For another, as earlier noted, it is spatial practice that transforms a place into space. Nevertheless, tourist activities are remarkably diverse. Besides, this research is not a typical tourist experience oriented tourism study, but mass communication-centered. In view of the difficulty and the specific purpose, my strategy for analyzing the historical sites is to investigate one of the key stakeholders of touring, the tour guide. Drawing from my fieldwork, I analyze the walking tour of the Yangjialing site. It should be noted that the qualitative data being used in this section were compiled from multiple tours that I attended, not just from one. Also, all the tour guides' names are pseudonyms.

Yangjialing is about 2 kilometers north of Yan'an city. Due to Japanese bombing, the CPC Central Committee (CPCCC) along with its headquarters moved to Yangjialing from Yan'an in November 1938 and stayed until March 1947. The site was where many of the party's defining moments occurred, noticeably, the Yan'an Rectification and the Yan'an Talks. The small mountainous village of Yangjialing is presently occupied by historical buildings of the CCPCC, including the Central Auditorium, major bureaus and offices, and the residences of the CPC's early leaders. In addition to the Auditorium, tourists come to Yangjialing mainly to see the *yaodong* caves where Mao Zedong, Zhou En Lai, Zhu De, Liu Shaoqi, and other leaders once resided. Tourists are grouped for a guided walking tour either by the on-site tour guide or by the regular tour escort/guide.

A fundamental form of spatial practice, walking is an intermediate state. That is, walking is liminal (Skinner, 2016; Solnit, 2002). It involves ambiguity, orientation, disorientation, and reorientation with regard to reading. This is the reason why walker/walking is of paramount importance to de Certeau's (1984) thinking of the practice of everyday life and the strategy–tactic framework. In a walking tour setting, the spatial practice of walking is highly organized and guided by both the tour guide and carefully selected routes. Hence, the walking is nothing natural like a daily practice, so is the reading. It is a production of management of spatial fluidity and reconfiguration of historical narrative. In that sense, the reading of space by the tourist turns into a matter of indoctrination by the tour guide. But this is done in a fairly fascinating way, a fabulist way to be more specific.

> The internal display of the Auditorium was mounted according to the original setting of the Seventh National Congress of the CPC. Look at the central podium. Hung in there are the side portraits of Mao Zedong and Zhu De. Hung above is the Seventh Congress' political slogan, "Victory under the Banner of Mao Zedong." … Let's look back at the wall. The four characters "One Heart, One Mind," was the key motto of the Seventh Congress, which was handwritten by Mao Zedong. From April 23 to June 11, 1945, the Seventh National Congress of the Communist Party of China was solemnly held in this auditorium. In addition to the Congress, there were also other large-scale activities held in here such as the premiere of *The white-haired girl* (白毛女).[4] On November 30, 1946, Commander-in-Chief Zhu De's 60th birthday was also held in the Auditorium. That's my introduction to the Central Auditorium. Now I give you a few minutes to take a look at the podium.
>
> (Alicia, July 14, 2016)

The walking tour starts in the Central Auditorium. It is an ideal tour start point since the Auditorium is not only the historical marker of the Yangjialing site but also the closest relic to the main entrance gate. Alicia, an on-site tour guide, unfolded a historical moment in an expressive tone like a narrator reading a poem. The Auditorium is empty and dark, immediately evoking a sense of nostalgia. While the first part of the narration is a standard practice in tourism, the second part about the events of the show and the birthday party transforms the Auditorium from a static *place* of the Party's Congress to a dynamic *space* as lived through with music, applause, and laughter. Knowing that each tour in the era of digital mobile devices is necessarily a photographic tour, Alicia left enough time for the tourists to take photos. In this regard, the Auditorium is an ideal place as tourists are allowed to stand in the podium to pose as a leader giving a lecture. Just watching such activity alone is so much fun.

The Office Building of the CPCCC is the first stop of the tour and the last stop before climbing the mountain. In the three-story building, only the west wing is open to visitors. Inside is a dining hall, which doubled as a conference room. Most significantly, it was the place where Mao gave the Talks on literature and art.

> Look at the first picture on the wall in which teacher Wang Dahuan and the first-year student Li Bo of Lu Xun Academy of Arts performed *The brother and the sister reclaiming wasteland*[5] in Yan'an street. In the absence of

any PA device, the audience reached more than 20,000. You can tell from the photo, how immensely the masses were enthusiastic about this [performance]. Now let's look at [a giant photo of the scores of the song] *The East Is Red (dongfang hong* 东方红*)*. It was adapted from a traditional love song by a liberated peasant named Li Youyuan, who expressed the love of the people of northern Shaanxi for the Communist Party and the people's leaders. Now look at the next photo. This is Chairman Mao watching the Yang-ge Dance in the Spring Festival. Taken in 1943, this photo is the one that Chairman Mao's smile looks most cheerful, brightest, and happiest.

(Alicia, July 14, 2016)

Until now it is evident that a memory of Yangjialing is being rekindled by a series of iconic performances, rather than by a retelling of the grand narrative. Piecing these representations of performances together, the Yan'an Spirit, the representation space of revolutionary Yan'an comes to the fore. But this iconological style of narration changes noticeably after the tourists were taken to the mountain.

All former residences of the party's leaders are located in the loess mountain. These are *yaodong* caves, a representational space of Yan'an. Usually tour guides would take tourists to Zhou Enlai's *yaodong* first since it sits near the first turn of the road.

Li Peng[6] was already fifteen years old when he was in this place. (Then the tour guide points to the photo on the wall). The taller one is Li Peng, Premier Zhou's adopted son. The next two tile-roofed houses were the places where a lot of orphans of martyrs lived. They were adopted by Deng Yingchao (Mrs. Zhou Enlai). Li Peng was the son of Li Shuxun. The orphans grew up in a nursery, including (Gen.) Liu Bocheng's (adopted) son Liu Taihang. This photo shows that Premier Zhou got injured. He fell from a horse, which is the reason why his arm could only bend sixty degrees. This is the famous story that was later called "Falling from a Horse into the Yanhe River."

(Rachel, July 12, 2016)

Unlike Alicia, Rachel is a tour escort from a southern province. When visiting Zhou Enlai's residence, Rachel spoke of Li Peng with only a passing remark on an incident of Zhou Enlai. Nevertheless, Rachel only conjured up an image of the controversial figure without going any further and left the rest to the tourists' imagination. The same narrating style also happened in Mao Zedong's residence, the most crowded place of the Yangjialing site. This time the controversy centers on Mao's fourth wife and their daughter.

The third cave was the bedroom of Chairman Mao and Jiang Qing.[7] See the earth cave on the top? It was the place where Li Na, the daughter of the Chairman and Jiang Qing, was born in August 1940. Li Na lived in the cave with her nanny alone since her birth. Li Na comes here at least twice a year. Li Na will not go anywhere but directly to here off the plane in the airport.

When she left here, she was already eight years old, so she had an affectionate memory of this place.

(Mark, July 13, 2016)

Mao Zedong and Jiang Qing married in Fenghuang Mountain on November 20, 1938. They moved to Yangjialing the same day because the Japanese bombed Yan'an. So, their wedding was in Yan'an but the bridal chamber was in Yangjialing. A foreign journalist who came to China said emotionally, "the party's leader wrote for a long time in such a cold cave in the faint light. There were no exquisite furnishings, no material enjoyment. But the man who lived in the cave was sharp, thoughtful, and had a vision of the world. That man is Mao Zedong." Well, take a look at the bridal chamber of the Chairman and Jiang Qing.

(Amy, July 13, 2016)

Both Mark and Amy are tour escorts. Like all other tour guides, they highlighted the enigmatic intimacy between Mao Zedong and Jiang Qing alongside their daughter in narrating this site. While they did mention that the cave was also the birthplace of many of Mao's famous articles, the tourists seemed to show no interest in that passing remark. Throughout the walking tour, tourists asked most questions in front of Mao's cave. Rarely did the tour guide receive criticism or challenging questions from tour members. More often than not, the tour guide would get questions about the romantic relationship between the couple. No matter how trite those questions may sound, the tour guide usually sailed through with an authoritative tone. In addition to asking tourists to have a peek at the empty cave, Mark also cued them to rent the Chinese Red Army's costumes from a rental booth in the courtyard and to take photos in front of Mao's "house."

It should be noted further that as a major spot of the Yangjialing site, the display of Mao's cave does not tend to glorify Mao as a great political leader as one may assume. Featuring a bookcase of books, the exhibition represents Mao as a typical traditional Chinese literato with a passion for books and reading.

So why did the Chairman name his daughter Li Ne 李讷? This is, "a gentleman should be slow in assertion, but quick in action" [*junzi yu ne yu yan, er ming yu xing* 君子欲讷于言，而敏于行, from *The analects of Confucius*]. Ne (讷) means slowness. The Chairman wanted his daughter to talk less and do more. But later, many people misunderstood him, calling his daughter Li Na 李娜 instead. In fact, her name is Li Ne. Take a look at the bed where the Chairman slept. He had a habit of reading books and newspapers before going to bed. Therefore, he used the bookcase as the bedside table in each of his houses.

(Ted, June 10, 2021)

The intellectual image of Mao has been consistent in his media representations in post-Mao China. What is really surprising of the display, however, is to reframe

Mao as an ordinary person even being somewhat wayward. This is evident in many narrations of tour escorts:

> Mao Zedong was more than 20 years older than Jiang Qing. Jiang Qing arrived in Yan'an in 1937. Coincidentally, we all know that was the same year when He Zizhen [Mao's wife] left Yan'an. The reason why He Zizhen left Yan'an is very simple. It is a fact that can now be found in [historical] materials. Chairman Mao advocated learning from foreign countries and then, ballroom dancing caused a trouble. He Zizhen was unhappy about Chairman Mao learning ballroom dancing [without her], criticizing Mao about this in public. Mao was angry. So was He Zizhen. She left Yan'an for Xi'an without notice. Premier Zhou repeatedly persuaded her to come back, but she would not. Mao Zedong was determined to be marrying Jiang Qing. In light of such a helpless situation, the comrades had to acquiesce in the marriage. But the comrades set out three rules [for Jiang Qing]. [First,] Jiang Qing could only be responsible for the Chairman's daily life, somehow like a nanny looking after a person. [Second,] she was not allowed to go into politics. [Third,] her image could only appear in a family photo with the Chairman, absolutely not in any other work-related photos.
>
> (Tim, June 10, 2021)

The next stop is the former residence of Liu Shaoqi.[8] Tour guides commonly called it a "five-star hotel," unanimously referring to it as the equivalent of today's Diaoyutai State Guesthouse in Beijing. However, these are regular *yaodong* caves except for being used to accommodate generals and the top leaders after they returned from the battlefront. In fact, Liu Shaoqi did not have his own cave in Yan'an since he worked in the KMT-controlled area at the time. The cave was allocated as the "residence" of Liu Shaoqi in 1953 during the renovation of the Yangjialing site.

Chris is a Yan'an-based tour escort, quite knowledgeable about local folklore. Chris takes tourists to a small stone table near the busy road within sight of the caves. More likely tourists would pass by without noticing the table if they were not in a walking tour thanks to the glaring banality of the setting. However, the site/sight suddenly turns into a very dramatic one as tourists listen to Chris. Here I present at length below:

> American journalist [Anna Louise] Strong interviewed Mao Zedong right here. Chairman Mao said humorously, "all the US-Chiang reactionaries are paper tigers." Then Lu Dingyi,[9] sitting on the other side, translated "paper tigers" into "straw men." Chairman Mao shook his head, suggesting that the translation was not correct. Lu Dingyi was very smart and he responded very quick, correcting the translation as "both straw men and paper tigers." Chairman Mao then was pleased. He laughed, saying, "straw men can be put in the fields to protect crops from sparrows. If you leave a paper tiger in the fields, it can be easily destroyed by a rain, so the paper tiger is not as good as the straw man. And it (paper tiger) exactly represents the situation."

According to the sitting custom of our northern people, Chairman Mao must sit here, Strong there. There were a lot of older people and children standing in the mountain to watch this. When Strong turned around and found this, she asked why there were no people behind Mao Zedong. Chairman Mao laughed, saying, "these are my friendly neighbors. They have never seen anyone like you with blonde hair, high nose, and big eyes." Strong said to Chairman Mao, "well, I can change the position with you, so your friendly neighbors will have a good look at me, a foreigner." Then she stood here, saying *ha-lou* (hello) to the elderly and children in the mountain. They couldn't understand what she said. Finally, someone stood out, saying, "she asked us to go down." So, all the people ran down from the mountain. This is because "*ha-lou*" in the dialect of northern Shaanxi means "down."

(Chris, July 15, 2016)

In the course of his recounting, Chris imitates Mao's and Strong's demeanor and posture like an actor performing on the stage. Following Chris' cue, tourists light-heartedly and repeatedly speak "*ha-lou*" to imitate the funny accent.

The walking tour winds down on a hillside. It is a perfect end. From here tourists can take a rest and see a small walled vegetable plot down the road. That is the famous "Chairman Mao's Vegetable Plot." The Plot was featured in an article entitled "A morning in Yangjialing" that was included in the primary school textbooks from the 1960s to the 1970s and was standard reading for students from my time. Reportedly reclaimed by Mao himself, the plot was a signifier for the self-reliance tradition of the party.

The Chairman invited patriotic overseas Chinese Chen Jiageng 陈嘉庚 (1874–1961) for a dinner here using vegetables picked up from his own vegetable field. The dinner cost twenty cents, which was paid for a chicken brought by his neighbor. Chiang Kai-shek, too, invited Chen Jiageng for a dinner in Chongqing in order to fawn over him. The cost was eight-hundred yuan. Having his appetite fully satisfied, however, Chen Jiageng criticized Chiang Kai-shek (in Chongqing), "how the hell have you squandered such a great amount of money for a dinner! I donate money not for you to waste!" After visiting Yan'an, Chen Jiageng said, "the country will be the Communist Party's country!" Later, he donated a lot of money to the Communists.

(Chris, July 15, 2016)

Rather than proceeding to other sites on their own, tourists usually opt to go down and return to their tour buses. Walking tours have been so popular that sites excluded from the tours have become deserted. I had a keen interest in visiting the CPC's Propaganda Department in the Yangjialing site but all my attempts failed as the office seemed to have closed without notice.

The self-styled tour guides play a crucial role in tourists' imagination of the revolutionary heritage. Unlike the on-site narrator, the tour escort/guide wears ordinary clothes, speaks Mandarin but usually with an accent, thick or slight, a sign that the role did not require intense professional training. Nevertheless, this can be a strength for their storytelling. As escorts, they tend to cater to tourist groups creatively rather than performing a stiff recital. What's more, they stick to a conversational style and their accounts, though largely fabulist, are much more entertaining compared to on-site official narrators.

The entertaining style of the tour guides in Red Tourism sites sparks off continuing controversy. Referring to the phenomenon as "vulgarization and discoloration," and "distortion of history," the Red Office has routinely criticized this style in its annual reports, pointing out that the tour guides may confuse tourists when their interpretations deviate from that of on-site narrators (National Red Tourism Work Coordination Group (NRTWCG), 2005). Government authorities have further called for rigorous checks on both narrators and tour guides for a "comprehensive, accurate, and objective" interpretation of the history (NRTWCG, 2008). As an effort to establish and spread out exemplar models for tour guide narration, the National Red Tourism Tour Guide Competition, sponsored by the Red Office and based on CCTV and its online platform, had been held five times by 2014.

Whether entertaining or not, the walking tour is significant to both tourist experience and the social space of Red Tourism. Red Tourism sites like Yangjialing do not have many attractions upon which a tourist can gaze except for few historical buildings, caves, and photos. Historical artifacts are extremely rare, even replicas. Take Mao's cave for instance. Among three small, whitewashed rooms, there is Mao's bedroom. It is staged to be as if it were in the Yan'an days. The tourist can count the items in the display exactly on the fingers of one hand: a thermos bottle, a deckchair, a wooden bathtub, a bed, a bookcase, and a family portrait of Mao on the wall. This is to suggest that articulation of this space, to a great extent, depends on spatial practice characterized by the tour guide's narration in the movement.

Throughout the walking tour, guides develop pertinent tactics to cue, intrigue, and propel tourists into an imagined Xanadu, an idyllic and beautiful place where people were happy, singing, and dancing, and the leaders were amiable, eating vegetables they grew themselves. While holy sites are only for religious people, Xanadu is for everyone. One thing being omitted in the walking tour is the historical reality that the Yan'an period represents one of the most difficult times of the Party. This is why the Rectification, the Talks, the Spirit came into being in this space during this particularly adverse period.

The Fantasy *show: in light of being "revolutionary"*

The marriage of history and entertainment in Red Tourism is palpable. This is, however, quite understandable, considering that tourism itself is a recreation industry. In the Red Tourism industry, history, or revolutionary history to be more precise, has been commodified into consumer goods made readily ingestible for tourists via diversified performance and multimedia activities. I will elaborate this aspect of Red Tourism in the next chapter within a new framework that I refer to as the commodification of propaganda. Suffice to say at this juncture that the commodification of the revolutionary past offers the tourists what Hannigan

(1998) would call a "participatory fantasy experience" (p. 26). As a result, the formerly sacrosanct boundaries – between dull political propaganda and fan-chasing tourism, and between political persuasion and business maneuver – have been blurred, collapsed, and recast into something fantastically new. Take, for instance, the *Fantasy of the Yan'an Defense*, an on-site reenactment of a historic battle.

Branded as "A Must See for Any Tourist in Yan'an," marketed with the gimmick of "real guns and real bullets," and audience participation, the *Fantasy* show was believed to be "revolutionary in Red Tourism" (Xu, 2010, p.204). It has now become one of the most popular sites of Yan'an tourism. According to a local news report, the outdoor show earned more than 2 million yuan ($320,000) within a few days during the Golden Week holiday of 2014 (Liu & Gao, 2014). The show depicts the Yan'an Defense as a grand spectacle where historical fidelity surrenders to vaudevillian performance. The show incorporates many vernacular, traditional, and representational styles of dance and music.

The "battlefield" of the *Fantasy* is conveniently located at the foot of a loess hill a few kilometers north of the Zaoyuan 枣园 site. Usually performed twice a day, the show was only held once around noon in summer 2016 when I attended. Tour buses arrived right before the show started. Somewhat disappointing is that the battlefield is merely composed of several rows of crude bungalows. The setting did not convey the intensity of war. I sat on the bleachers facing a vast open field along with about six hundred tourists as if we were going to watch a football game except no athletes, only performers.

The show starts with a very peaceful, lovely day in the Yan'an Period. Communist troops train in the front. A group of women soldiers sit on the ground spinning in the back against another group of male farmers and soldiers plowing at the far end. A visual representation of self-reliance and the Yan'an Spirit, the opening scene alongside the whole setting is distinctly familiar to the audience as it has been a staple of Yan'an-themed television dramas and films since the dawn of the Red industry. Then a high-pitched suona sound was heard. A double-reeded horn, the traditional Chinese music instrument suona is usually performed at rituals. Immediately, the cheerful music signals a traditional wedding ceremony underway. The bridal retinue comes from a distance with the bride riding a donkey. The wedding ceremony reaches its crescendo as the Communist soldiers join in. Even until this moment my disappointment was not assuaged as the reenactment was merely a musical, and there was no sign of what the advertisement has billed as a "real war." But I was proven wrong.

The first explosion sounded just like a real attack. A signal of the beginning of the Yan'an Defense, the explosion was loud enough to diminish my hearing. More explosions occurred shortly as the KMT air strikes began. I heard children crying in the audience. Choking and pungent smoke filled the air in no time. A few miniature WWII fighters were moving, swiftly along on suspended cables and firing guns. The happy wedding soon turned into a tragedy as people died. More troops from both the Communist side and the KMT entered the scene. I could hear, however, only the buzzing, ringing, and whistling sounds. The audience was as excited and attentive as if they were in a real battle.

The bomb blasts were ridiculously realistic. It is simply because they came from real bombs and explosives. Now I believed the show's hype. Even worse, I could barely breathe due to the overwhelming heat from the explosions. I was the nearest spectator to the bombs. While all the audience were relatively close to the explosions, I sat at the right end of the first row of the bleachers for better photography. The nearest explosion was no more than a road-breadth away from me. And on that road were real tanks firing and bombs setting off. I could feel, from the overwhelming heat and smell, that the explosives were mixed with gasoline. Perhaps, this is a stage trick for adding more special effects to an explosion. But my feeling that someone can easily and seriously get hurt was by no means an exaggeration. I have learned that in 2010, an actor was accidentally shot by a gun in the same show. Elsewhere, several staff members were killed during pyrotechnics setup for a similar show produced by the same company. In that sense, the *Fantasy* is fantastically eye-opening and breathtaking regardless of its narrative.

The show ends up with the red flags of the Communist troops being raised in Yan'an city. Echoing the ending scene, more flags are being waved in the loess hill. To highlight the victory further, hundreds of men and women start up the Ansai Waist Drum Dance in the field. A representation of the space of Yan'an, the same large-scale drum performance featured in the overture of the 2008 Beijing Olympics opening ceremony. The drums continued roaring, as did the suonas as if the show would never end. Shortly, performers start to invite the audience to join the dance. Early on, other tourists wore the costumes and participated in the show as extras (background actors) by paying extra. At the end, tourists took pictures with the performers and lingered around. To me, the show is well worth attending for the shock and awe rarely seen in other outdoor theaters. But in the interest of personal safety, perhaps once is enough.

Tom, a tourist from Beijing, is keenly aware of the "fantastic" flavor of the *Fantasy*. He says:

> The show was interesting and I had so much fun … I thought it would be really fun to participate in the show, so I paid extra for the interactive session. My only regret is that I got the tattered [prop] pants. I wish I could have played the Kuomintang. In that case, I could at least wear a uniform. Our tour guide told us that it is a live show with real ammunition. But I feel like I watched folk dances and listened to folk songs for the most of the time. The real deal of the show is the beautiful spy; she became especially attractive when riding a horse.
>
> (interview, June 26, 2013)

To Tom, the show is simple entertainment, no more special than any other commercial performance. As long as the show was entertaining, Tom did not seem to care much about either history or the revolution.

The *Fantasy* is a commercialized space. Within that space, a piece of revolutionary history is not so much being reenacted, represented, and proliferated as being simply commodified. This is unsurprising, considering that the show was an investment by a single entrepreneur from Wenzhou, a southeastern city of China known as the paradise of private enterprise. The total investment in the show reached 10 million yuan ($1.6 million) including enlisting Chen Weiya, a famous director and choreographer who codirected the spectacular opening ceremony of the Beijing Olympic Games (YTB, n. d.). There is also an official website created for the show. Despite the show's entertainment value, this entrepreneurial deployment of "Red" seems to be a cross-border appropriation of the post-socialist sentiment and a resignification of "Red" in contemporary China. As this book was being completed, the show changed its early name by dropping off the word "fantasy," now it is called the *Yan'an Defense*. By 2021, the overwhelming majority, if not all, of audiences of the reenactment show were organized tourists by tour companies.

Post-socialist nostalgia, stress, and emergent counternarrative

The success of Yan'an Red Tourism, however, raises a few critical questions that cannot be explained by spatial practice. For example, why has Red Tourism increasingly gained popularity at a time when market capitalism, rather than prior forms of socialism, is prevailing in China? Is this Red pilgrimage a result of state propaganda or a reflection of social demands unfulfilled by the post-socialist state? There are no easy answers since where Red Tourism goes remains to be seen. But there are clues.

Red Tourism offers relief on the one hand to the feelings of loss residing in the Chinese public's nostalgia and imaginations of the socialist past, as well as anxieties and social stress brought on by market economy. Whyte (2012) regarded the feeling of post-socialist stress in China largely as a response to briberies and embezzlements, and this is the general impression that I gained from many inter-viewees as well. Susan is an official from the local tourism bureau. When asked about the meaning of Yan'an Red Tourism, she says:

> Currently, the party calls for spreading positive energy. All stories of Yan'an are positive energy. Looking at those corrupted officials and then looking back at the old generation of the party officials and leaders who worked here, it is quite obvious that they are nothing like those [today's] officials; they stuck to plain living. Even high-level officials at that time were quite approachable by ordinary people.
>
> (interview, June 25, 2013)

Susan expresses her dissatisfaction with the reality in contrast with the past glory of the Mao era. That dissatisfaction did not come from a vacuum.

China's economic reforms of the1980s resulted in some eerily new Chinese characteristics, as Harvey (2005) put it, describing "a particular kind of market economy that increasingly incorporates neoliberal elements interdigitated with authoritarian centralized control" (p. 120). Zhang Xudong has also gotten a strong "mixed-ingredients" flavor, terming it "postsocialism." Specifically, Zhang

(2008) argued that postsocialism is "a conceptual proposal to stay and live in contradictions and chaos in a mixed economy and its overlapping political and cultural (dis)order" and also, "a result of the historical overlap between the socialist state-form and the era of capitalist globalization" (pp. 15–16). As Jameson (2000) observed, one far-reaching consequence of neoliberal globalization is that "culture has become decidedly economic" (p. 54). To the extent that this shift came largely unexpected and with astonishing speed, it is reminiscent of what Charlie Chaplin faces in front of the assembly line in *Modern times*, except this time it was not speeding conveyor belt from the industrial revolution but a radical change in the culture industry. Nevertheless, the feelings of anxiety, loss, and longing nestling in nostalgia are strikingly similar.

In reconceptualizing the term "nostalgia," Hutcheon (2000) valorized the emotional impact of the pastness of the past, pointing out that the past is not something experienced but imagined. Since such imagination is an ongoing process, Hutcheon (2000) argued that "nostalgia is less about the past than about the present," or it "exiles us from the present" to be more precise (p. 195). Perhaps this can, in part, explain why Red tourists, by simple observation, are not exclusively the elderly but people of all ages. Dai (1997) attributed the nostalgic sentiment of the 1990s in China to rapid urbanization, calling it an "imagined heaven" (p. 148). To build that heaven, the cultural industry started with exploiting nostalgia. Dai (1997) remarked that as part of marketization of nostalgia, revolutionary history was sexually romanticized in popular culture.

It should be noted that the revolutionary past has been a key historical object for nostalgic appropriation in Chinese cultural industries for decades (Wu, 2006). Coincident with the CPC's efforts to reinforce the socialist value system in the 1990s, Red Tourism benefited from post-socialist nostalgia. Many of the interviewees indicated that visiting Yan'an had fulfilled their long-held dream, nostalgia for the Mao era. Richard, a tourist, says:

> My family drove all the way to Yan'an from Wulumuqi.[10] I like history. I have been longing for Yan'an since I was a kid, because it is the sacred place. I was particularly amazed by He Jingzhi's *Return to Yan'an*[11] when I was young ... It [Yan'an] provides the Chinese people with great spiritual wealth. I think my biggest gain of this trip is that our kids get to know how hard the revolution was and the life of the great revolutionaries, all of which would be very valuable for their growth.
>
> (interview, June 23, 2013)

In his early forties, Richard did not experience any revolutionary moment associated with Yan'an, but he has a dream about the revolution. And he wants his children to be inspired by what he believes what Yan'an represents.

Another clue is held in an emergent counternarrative. The innermost part of Red Tourism is the grand narrative. As shown earlier in the RMY, the grand narrative presents to the tourists a kind of emancipation narrative. As such, it links the revolutionary past to the post-socialist present in a predetermined way insofar as each historical event and figure was assigned a value and a character to transform history into a saga from which the Chinese nation-state and its identity are constructed and propagated. To the extent that Red Tourism is part of a long-lasting ideological education campaign, it is fair to say that the saga or the grand narrative is more about casting light upon the present than articulating the past. In Lyotard's (1984) original account, the grand narrative matters to social bond. In the postmodern condition, for example, the collapse of the grand narrative led to social disintegration and social atomization in an erratic random movement, which Lyotard (1984) referred to as "Brownian motion" (p. 15). Red Tourism as a social space of propaganda, or a force of social integration, is meant to counter that effect and the corresponding movement.

A counternarrative emerges from the Red tourist's experience. The fact that history museums encourage visitors to engage in intended imaginations also suggests that they inhibit the visitors from imagining alternatives (Wallace, 1996). But imagination, perhaps, is one of the hardest things to control. A fantastic lion circus show for some people may stir an imagination of animal cruelty for others. In a Red Tourism setting, the touristic gaze is set in a dynamic history-reality comparative structure from which tourists' interpretations and imaginations arise. The comparative structure can be seen as an "internal storyline," to borrow a term from Doering and Pekarik (1996). The tourist brings it to the Red Tourism site as the foundation for interpreting the display. Specifically, Red tourists interpret museum displays and form their own opinions by comparing historical practice to the current one. Typically, this process likely results in their dissatisfaction with the present. It is this juxtaposition of the lyrical and poetic revolutionary past as mediated in exhibits, tours, and shows with the bitter reality that creates a counternarrative running through Red tourists' experience.

The counternarrative serves as a timely reality check for the tourist. Two aforementioned stories of the RMY illustrate this point. Supposed to propagate a democratic and law-abiding image of the party, however, the stories about the beans and the general-*cum*-murderer oftentimes invoked tourists' bitter resentment of reality. In light of the democratic revolutionary days, what the interviewees spoke most of was the current corruption and graft in officialdom. Jason, a cadre attending a training program in Yan'an, says:

> Although we have received "red" education from schools to work units and somehow become indifferent to it, we were still shocked in the RMY, particularly when we saw [the photo of] "Voting with Beans." Facing a lot of illiterates, our party still insisted on voting by secret ballot. The first day we arrived at Yan'an, we were invited for a welcome dinner by a unit. We drank fifteen bottles of wine worth 380 yuan ($57) each, and of course, it was not paid by ourselves but the work unit. I knew it was not right but you know, when in Rome, do as the Romans do. Had we stuck to only half principles of what the early Party members did, the officials would not have corrupted.
>
> (interview, July 22, 2015)

This counternarrative has been consistent and pervasive throughout my interactions with tourists, tour guides, and government officials, whether in formal interviews, informal chats, or overheard conversations. Beyond the discourse of revolutionary history, Red Tourism also includes a prominent and ongoing discourse of contemporary Chinese politics. This is an example of what I have noted earlier that a mobile Chinese public sphere can emerge from a propaganda sphere. Gary, a senior CPC member who takes a walking tour of the Yangjialing site with other party members, asks:

> Why is there a sign of more corruption occurring even in the course of China's massive anti-corruption campaign? Why cannot this campaign achieve a similar effect as the Yan'an Rectification Movement? The Rectification took the mass line, whereas today's anti-corruption campaign takes a different pathway.
>
> (interview, July 11, 2014)

The emerging counternarrative, however, does not imply that the grand narrative has fallen apart. Not even a sign. Red tourists, to a great extent, cherish the memory of the authentic revolutionary past. This is why family groups are a frequent presence at Red Tourism sites. Parents bring their children to the site for the purpose of education, not specifically for a history but for the purpose of instilling the parent's own worldview and values. Their practice can be seen as a form of *voluntary* indoctrination as the morals and values of revolutionary heritage are inculcated in children subtly, but entirely deliberately. On-site education by parents takes many forms. Parents may directly talk to their sons and daughters to draw life lessons from the early revolutionaries. Sometimes, kids take hints from their parents to gaze on particular things supposed to be inspirational for their growth. Kelly, a young mother visiting the RMY with her husband and their 6-year-old son, says:

> For us, Red tourism is merely about relooking at a history, but for our children it is more than that. Now most children are the only child of their families. They do not care about other people. It seems that everyone else lives only for them. They can easily get what they want. Although our kid does not understand history, we brought him here, at least, to let him see a lot of people who lived for others. Also, we wanted to let him to learn good things such as hard work attitude and simple life style. I think these are things that he cannot learn from other kinds of touristic places.
>
> (interview, July 22, 2015)

The enduring grand narrative together with the emergent counternarrative is a hint at tourists' ambivalence towards Red Tourism. Their interpretations and imaginations during the Red tour are dynamic, undefined, and always subject to reconstruction with a discourse of current politics being dragged into the historical discourse.

Conclusion

The tourists' ambivalence points to a social space of Red Tourism. As such, Red Tourism is not determined by propaganda, and not determining such propaganda either.

The Yan'an case illustrates the polyvalence of the social space of Red Tourism. On the one hand, Red Tourism is a space produced by revolutionary history, the current party's politics, post-socialist stress, and a market economy, among others. On the other hand, Red Tourism is producing a kind of hybrid modernity, new urbanscapes, a Red economy, and a counternarrative in addition to the grand narrative.

In the (re)production of the revolutionary past, space always matters. Central to the city's new gentrification and resettlement plans is the Red Tourism orientation. The Red-themed environment of Yan'an led Australian scholar Roland Boer (2013) to exclaim in his travel blog, "China may have communist symbols throughout, but compared to Yan'an, they seem sparse indeed. Every bridge, every building, every road, every poster proudly displays red flags, red stars, hammers- and sickles, and what have you." On the micro level, commercial reenactment shows like the *Fantasy* offer Red tourists an escapist, almost daydreaming experience in an extraordinary way as promised by the show's promotion. The fantasy goes on to the walking tour of Yangjialing. In contrast to the deafening spectacle of the *Fantasy* show, an image of tranquil Xanadu emerges from the tour guides' narrations. While the RMY holds considerable space-oriented institutional power in delivering the grand narrative, the visitors, through spatial practice-oriented "tactics," construct a counternarrative with which they not only make sense of the past, but more importantly, formulate a critical view about reality in an ambivalent way where the grand narrative coexists with the counternarrative. This can be a manifestation of what a mobile Chinese public sphere looks like.

The social space of Red Tourism has evolved and expanded massively as Red Tourism transformed from a small business on the town and village level to a capital-intensive regional industry. Pertaining to this sea change, the emerging power relations between the state and the capitalist in the production of Red Tourism, the private channel for distributing institutional power, labor issues, and the capitalist production model render classical theories of tourism, propaganda, and semiotics helplessly insufficient. Thus, a political economy of Red Tourism becomes imperative. This is what I will set out to formulate in the next chapter.

Notes

1 The Yan'an Talks was included in Mao's (1965) *Selected Works of Mao Zedong, Vol. III.* Beijing: Foreign Language Press. The English translation of the full text can be retrieved from https://www.marxists.org/reference/archive/mao/selected-works/volume-3/mswv3_08.htm

2 Three mountains refer to Pagoda Mountain (宝塔山), Fenghuang Mountain (凤凰山), and Qingliang Mountain (清凉山), whereas two rivers refer to the Yanhe River (延河) and the Nanchuanhe River (南川河).

3 In this article Peaslee (2013) defines "media conduction" as "movement of information due to a difference in level of access (from a high access to a lower-access region) through a transmission medium (e.g., festivals, conventions, events) that simultaneously reifies the value of that access" (p.811).

4 The *white-haired girl* (白毛女) is one of the Chinese classic revolutionary operas, later adapted to film, Beijing Opera, and ballet. The opera is based on a folk legend circulating in the Border Region of Shaanxi, Chahar, and Hebei, describing the life change of the peasant girl Xi'er before and after joining the communist force. A Propaganda work encapsulating class struggle, the show quickly became a marker and thus an embodiment of the Yan'an Talks after its premiere in 1945.

5 The *brother and the sister Reclaiming Wasteland* (兄妹开荒) is a short musical created in the Yan'an Period. Based on local music and the Yang-ge Dance (秧歌舞), the musical captures a few moments of the Large-scale Production Movement (*da shengchan yundong* 大生产运动). It soon became of an iconic representation of self-reliance.

6 Li Peng 李鹏 was former Premier of the PRC (1987–1998).

7 Jiang Qing 江青 (1914–1991) was a Shanghai-based actress before coming to Yan'an in 1937. A member of the "Gang of Four," Jiang Qing was a key figure in launching the Cultural Revolution in 1966.

8 Liu Shaoqi 刘少奇 (1898–1969) was a leading statesman, Vice Chairman of the CPC (1956–1966), and President of the PRC (1959–1968). Labeled as a traitor to the revolution and an antagonist of Mao Zedong, Liu Shaoqi died under harsh treatment in the early stage of the Cultural Revolution.

9 Lu Dingyi 陆定一 (1906–1996) was the editor-in-chief of the *Liberation Daily* (解放日报) at the time. He served as the first director of the Propaganda Department after the founding of the PRC.

10 Wulumuqi is the capital of Xinjiang Uyghur Autonomous Region in the northwest of China.

11 He Jingzhi 贺敬之 is a Chinese poet and playwright. He went to Yan'an in 1940 when he was 16 and joined the Party a year later. During those days, He Jingzhi wrote his masterpiece, *Return to Yan'an*, a poem in which the poet affectionately calls Yan'an "mother." The poem was included in many editions of primary and secondary school textbooks.

References

Anderson, B. (2006). *Imagined communities: Reflections on the origin and spread of nationalism*. Verso.

Apter, D. E. (1993). Yan'an and the narrative reconstruction of reality. *Daedalus, 122*(2), 207–232.

Benjamin, W. (1979). *One-way street and other writings*. Verso.

Bennett, T. (1995). *The birth of the museum*. Routledge.

Benton, G. (1975). Introduction to "the Yenan literary opposition". *New Left Review, 1*(92), 93–106.

Boer, R. (2013, December 10). The Yan'an spirit. https://voyagesontheleft.wordpress.com/2013/12/10/the-yanan-spirit/

Bourdieu, P. (1989). Social space and symbolic power. *Sociological Theory, 7*(1), 14–25.

Carey, J. W. (2009). *Communication as culture: Essays on media and society*. Routledge.

CPCCC & the State Council. (2011). *2011–2015 National Red Tourism development outline*. https://www.renrendoc.com/paper/97704257.html

Dai, J. (1997). Imagined nostalgia. *Boundary 2, 24*(3), 143–161.

de Certeau, M. (1984). *The practice of everyday life*. University of California Press.

Doering, Z. D., & Pekarik, A. J. (1996). Questioning the entrance narrative. *Journal of Museum Education, 21*(3), 20–23.

Eco, U. (1997). Function and sign: The semiotics of architecture. In N. Leach (Ed.), *Rethinking architecture: A reader in cultural theory* (pp. 182–201). Routledge.

Fan, N., & Hu, B. (2013). Yanan zhengfeng kaichuang liao dangnei minzhu jianshe de xin moshi [The Yan'an Rectification built a new model for the construction of inner-party democracy]. *Qingnian yu shehui, 27*, 1–4.

Goldman, M. (1967). *Literary dissent in Communist China.* Harvard University Press.

Guo, B. (2006). *Yan'an xue yanjiu diyi ji* [Yan'an studies, volume 1]. Zhonggong dangshi chubanshe.

Hannigan, J. (1998). *Fantasy city: Pleasure and profit in the postmodern metropolis.* Routledge.

Harvey, D. (2003). *Paris, capital of modernity.* Routledge.

Harvey, D. (2005). *A brief history of neoliberalism.* Oxford University Press.

Housing and UrbanRural Development Bureau of Yan'an. (2011). *The overview of the general plan of beautifying Yan'an, 2011–2015.* http://www.yanan.gov.cn/info/1241/60048.htm

Hutcheon, L. (2000). Irony, nostalgia, and the postmodern. In R. Vervliet & A. Estor (Eds.), *Methods for the study of literature as cultural memory* (pp. 189–207). Rodopi.

Irving, R. J. (2016). Implementation of Mao Zedong's Yan'an "Talks" in the Subei Base Area–The Chen Dengke "phenomenon." *Asian Studies Review, 40*(3), 360–376.

Jameson, F. (2000). Globalization and political strategy. *New Left Review, 4*, 49–68.

Judd, E. R. (1985). Prelude to the "Yan'an Talks": Problems in transforming a literary intelligentsia. *Modern China, 11*(3), 377–408.

Kracauer, S. (1995). *The mass ornament.* Harvard University Press.

Lefebvre, H. (1991). *The production of space.* Blackwell.

Leon, W., & Rosenzweig, R. (1989). *History museums in the United States: A critical assessment.* University of Illinois Press.

Li, F., Li, Z., Wang, Y., Li, F., Su, J., Luo, L., & Li, J. (2011). Chengshi tese yingjian—"Yan'an ba jing" gouxiang [Creating the features of the city: Thinking of the Eight Scenes of Yan'an]. *Anhui nongye kexue, 39*(16), 9956–9958.

Liu, X., & Gao, L. (2014, October 14). Yan'an baoweizhan jingqu guoqing jiaqi shouru chao erbai wan yuan [The Yan'an Defense scenic spot earned more than 2 million yuan during the National Day holidays]. *Yan'an Daily.* http://www.yadaily.com/News_View.asp?NewsID=26288

Liu, Y. (2006). *Shinian jishi 1937–1947 nian Mao Zedong zai Yan'an* [A ten-year chronicle: Mao Zedong in Yan'an, 1937–1947]. Zhonggong dangshi chubanshe.

Lu, X. (1999). Yan'an zhengfeng yu "sanjiang" jiaoyu [The Yan'an Rectification and the "Three Stresses" education]. *Henan shifan daxue xuebao, 26*(4), 90–92.

Luo, T. (2002). Lun Yan'an zhengfeng yu "sanjiang" jiaoyu [On Yan'an Rectification and the "Three Stresses" education]. *Gansu shehui kexue, 3*, 56–59.

Luo, W. (2012). Yan'an zhengfeng yundong yanjiu wenxian de tongji fenxi—yi zhongguo qikan wang pianming han "Yan'an zhengfeng" de wenxian wei fenxi duixiang [A statistical analysis of literature on the Yan'an Rectification Campaign: Based on the keyword search of "Yan'an Rectification" of Chinese journal articles]. *Fujian dangshi yuekan, 14*, 52–56.

Lyotard, J. F. (1984). *The postmodern condition: A report on knowledge.* University of Minnesota Press.

Massey, D. (1984). *Spatial divisions of labor.* Macmillan.

National Red Tourism Work Coordination Group. (2005). *Zhongguo hongse lvyou fazhan baodao 2005* [China Red Tourism development report, 2005]. Zhongguo lvyou chubanshe.

National Red Tourism Work Coordination Group. (2008). *Zhongguo hongse lvyou fazhan baogao 2007* [China Red Tourism development report, 2007]. Zhongguo lvyou chubanshe.

Nimkoff, M. F. (2008). *Media memories: The Newseum story of news* [Unpublished doctoral dissertation]. University of Illinois at Urbana-Champaign.

Peaslee, R. M. (2013). Media conduction: Festivals, networks, and boundaried spaces. *International Journal of Communication, 7*, 20.

Qing, S., Jiang, G., & Yao, X. (2014). Jiucheng shujie beijing xia de lishi wenhua yizhi baohu yanjiu—yi Yan'an hongse geming yizhiqun baohu weili [An examination of protection of historical sites in light of old city redesign: A case study of Yan'an revolutionary sites]. *Xiaochengzhen jianshe, 6*, 79–84.

Selden, M. (1995). *China in revolution: The Yenan way revisited.* M. E. Sharpe.

Seybolt, P. J. (1986). Terror and conformity: Counterespionage campaigns, rectification, and mass movements, 1942–1943. *Modern China, 12*(1), 39–73.

Skinner, J. (2016). Walking the Falls: Dark tourism and the significance of movement on the political tour of West Belfast. *Tourist Studies, 16*(1), 23–39.

Snow, E. (1938). *Red star over China.* Random House.

Solnit, R. (2002). *Wanderlust: A history of walking.* Verso.

Song, Z. (2001). Cong Yan'an zhengfeng dao "sanjiang" jiaoyu [From the Yan'an Rectification to the "Three Stresses" education]. *Dangshi yanjiu yu jiaoxue, 157*, 40–42.

Statistical Bureau of Yan'an City. (2016). The statistical communiqué of Yan'an city on the 2015 national economic and social development. http://www.yanan.gov.cn/info/egovinfo/info/Infor__con/016074151/2016-0016.htm

Wallace, M. (1996). *Mickey Mouse history and other essays on American memory.* Temple University Press.

Wang, K. (2014, November 11). Revolutionary justice. *China Daily.* http://usa.chinadaily.com.cn/epaper/2014-11/20/content_18950382.htm

Wang, K., & Wang, L. (2005, May 10). Yan'an "hongse lvyou" kaifa yu baohu ruhe bingju [On the development and protection of Red Tourism in Yan'an]. *The People's Daily,* p. 5.

Wang, Q. (2007). Yan'an zhengfeng yundong Zhong zhengzhi pipan moshi de jianli [The model of political criticism in the Yan'an Rectification Campaign]. *Xi'an wenli xueyuan xuebao, 10*(2), 44–47.

Wei, C. (1999). Cong Yan'an geming jinianguan de chenlie shijian tan zhanshi huanjing de kongjian sheji [The spatial design of the Revolutionary Museum of Yan'an in light of the practice of displaying]. *Zhongguo bowuguan, 4*, 67–70.

Wei, C. (2000). Cong Yan'an geming jinianguan de xuting sheji shuoqi [On the design of the entrance hall of the Revolutionary Museum of Yan'an]. *Zhongguo bowuguan, 2*, 51–53.

Whyte, M. K. (2012). China's post-socialist inequality. *Current History, 111*(746), 229–234.

Wu, J. (2006). Nostalgia as content creativity: Cultural industries and popular sentiment. *International Journal of Cultural Studies, 9*(3), 359–368.

Wylie, R. F. (1980). *The emergence of Maoism.* Stanford University Press.

Xi, J. (1999). Yan'an zhengfeng de lishi jingyan jiqi xianshi yiyi [The historical experience and practical significance of the Yan'an Rectification]. *Dangzheng luntan, 5*, 4–7.

Xu, R. (2010). *Zhongguo hongse lvyou yanjiu* [A study of Red Tourism in China]. Zhongguo jinrong chubanshe.

Yan'an jian ying [Introduction of Yan'an]. (1947, March 21). *Shen Bao*, p. 1.

Yan'an Municipal Government. (2012). Yan'an shi chengzhi zongti guihua, 2012–2030 [The master plan of Yan'an city, 2012–2030]. http://www.yasrd.gov.cn/Item/Show.asp ?m=1&d=3231

Yan'an Municipal Government. (2013, May 23). *Hongse lvyou ziyuan fengfu* [Red Tourism rich in resources]. http://www.yanan.gov.cn/info/1011/2037.htm

Yan'an Tourism Bureau. (2012). *The development of Red Tourism industry in the city of Yan'an*. Government document.

Yan'an Urban Planning and Design Institute. (2011). *Yan'an shi guihua shejiyuan 2011 niandu gongzuo zongjie* [Work summary of 2011 of Yan'an Urban Planning and Design Institute]   http://www.yajsgh.gov.cn/website/main/yadetailpage.aspx?fid=50eb6ca2 -42f3-4f12-84ad-368877ecd4a8

Zhang, B. (2011). Guanyu zuomei Yan'an de diaoyuan baogao [A research report on the Beautifying Yan'an project]. *Yan'an zhiye jishu xueyuan xuebao, 25*(4), 1–4.

Zhang, X. (2008). *Postsocialism and cultural politics: China in the last decade of the twentieth century*. Duke University Press.

Zhao, S. (2005). Yan'an jianghua yujing xia He Qifang wenxue guannian de gaizao [The transformation of He Qifang's attitude towards literature in the context of the Yan'an Talks]. *Wenyi lilun yanjiu, 4*, 10–18.

Zheng, P. (1999). Yan'an zhengfeng yu "sanjiang" jiaoyu [The Yan'an Rectification and the "Three Stresses" education. *Qiu Shi, 13*, 11–13.

Zhu, H. (2012). Yan'an gaibian liao zhongguo [Yan'an changed China]. In *Yan'an Narratives* (pp. 1–11). Zhonggong dangshi chubanshe.

6 The commodification of propaganda

Commodification is an inseparable part of the social space of Red Tourism. The acceleration of such commodification has been startling. In 2015, led by a couple of large corporations such as Shaanxi Tourism Group Co. and Shaanxi Culture Industry Investment Group Co., the total investment by private enterprises in Red Tourism in Yan'an and its proximity had reached 50 billion yuan ($7.7 billion) (Tian, 2014). Knowing that the total investment of Shanghai Disneyland is $5.44 billion (Palmeri & Lin, 2014), one cannot help wondering what kinds of Red Tourism projects those were. No grand theme parks or other spectacular man-made landscapes appeared. Added to the map of Yan'an Red Tourism were merely a few small dots that are negligible for their insignificance to the city's Red Tourism. Where did the money go?

In contrast to the Red Tourism industry as shown above and also in media coverage, complaints about the unprofitability of Red Tourism by local officials, tour guides, and tourism operators have been consistently revealed throughout research. It left many wondering whether Red Tourism is really a profitable business, and if so, then how does its business model work? This chapter addresses these conundrums by a political economy analysis of Red Tourism within a new framework, which I refer to as the "Commodification of Propaganda" (hereafter, CoP). CoP is understood as the production of propaganda in the private sector under the capitalist mode of production.

This chapter is based on my fieldwork in multiple sites in China conducted between 2011 and 2021. The main sources were key informant interviews selected based on their involvement in Red Tourism development over a long period. Given the propaganda topic of this study, the readers might be interested in knowing how I elicited management and leadership to speak candidly. There were no easy approaches and standard procedures. Data source triangulation was used for cross-checking information validity and capturing different dimensions of the conceptual construction of CoP. Among the respondents, three were senior officials of the tourism bureaus at three different levels, namely, provincial, metropolitan, and municipal; three were local government officials who were responsible for regional Red Tourism development; one was an entrepreneur ("capitalist") who conducted the "Red" reenactment show business in several Red Tourism sites;

DOI: 10.4324/9781003231783-6

five Red Tourism industry practitioners who provided me with insider views of the Red business; four workers who did different kinds of manual work in Red Tourism scenic spots. Added to the ethnographical data were news articles and commentaries pulled from both traditional newspapers and the Internet.

The following text is organized into three sections. The first lays the theoretical groundwork for later discussion of CoP. Rather than reiterating dialectical Marxism from ground zero, in this section I focus on a few interconnected terms, some latest developments within the political economy of communication in both Western and China's contexts, and land issues in China, all of which are aimed at providing theoretical and social contexts on which the CoP is based. The second section is to establish the CoP framework. Together, these are meant to provide theoretical pillars, social context, and an analytical tool that are necessary for an examination of the political economy of the Red Tourism industry. Composed of two cases, one reenactment business and one Red Tourism development project, the last section illustrates how CoP plays out in the real world.

Again, I herein adopt a neutral connotation of propaganda essentially equivalent to the traditional Chinese notion of *xuanchuan*, which refutes the popular assumption that all propaganda are necessarily misleading and deceptive.

Theoretical foundations and social context

In CoP, propaganda is conceived as some sort of raw material that is commodified for sale. This is not how Marxists commonly see propaganda (superstructure) and commodity (a good or service). Throughout the preceding chapters, I have argued and illustrated how propaganda as a social space is produced and productive, a move away from the traditional perception of propaganda. I now focus on commodity and related concepts and issues pertaining to CoP.

Commodification, commodity fetishism, commercialism, and hyper-commercialism

The commodity is the basis of the two most critical elements of Marx's conceptual framework: class and labor. Marx's analysis starts with the idea of the commodity (Harvey, 2010a). But propaganda, by nature, is not marketable; it must be commodified to be salable. The underside of such commodification is something Marx refers to as commodity fetishism.

In the original text, Marx does not elaborate on commodity fetishism, albeit being considered the "basis of Marx's entire economic system" (Rubin, 1973, p. 5) or an "essential tool for unravelling the mysteries of capitalist political economy" (Harvey, 2010b, p. 38). What's more, Marx's style in this section, as Harvey (2010a) pointed out, is fairly literary. Altogether, commodity fetishism leaves plenty of room for (re)interpretation.

The word "fetish" signifies two things: the inherent magical power of an inanimate object and worship towards that object by people. Marx (1967) uses the term "commodity fetishism" to address "the enigmatical character of the product of labor" (p. 82). Marx further explains the mysteries of commodities in that "the social character of men's labor appears to them as an objective character stamped upon the product of that labor" (p. 83). To put it another way, commodity fetishism is the mystification of the social relationships behind economic relationships among money and commodities exchanged in market trade. It transforms the subjective, abstract aspects of economic value into objective, real things that people believe have intrinsic value. With regard to Red Tourism, the social–political relations behind propaganda have been reconstructed through touristic experience as consumers of tourism; propaganda is still present, but reoriented from inculcation towards consumption. Having been commodified, the previously propaganda-relationship between the state and the masses was, too, transformed into one between the producer and the consumer, and accordingly, instillation becoming consumption, and finally, propaganda morphing into commodity. I will elaborate on this point shortly.

But commodification is not commercialization. The commercialization of propaganda in China and elsewhere in the world is not a recent development. Even in the height of the Cultural Revolution, production of propaganda goods for profit by private entrepreneurs did exist. Huang (2008) has remarked on a case: "[During the Cultural Revolution] another entrepreneur in the same township [Shishi] had raised 6,000 yuan from 36 investors and had started 30 small factories producing Mao Zedong pins" (p.62). While this anecdote illustrates how a businessman can take advantage of a political trend to make profit, it also suggests a chapter of capitalism pertaining to propaganda has yet to be written and seen. This is what I set out to do in this chapter.

The framing of CoP, however, is much more than a general notion of commercial use of propaganda. It is not about manufacturing Mao Zedong pins and other souvenirs with propaganda content and selling them to tourists for profit. Rather, CoP is meant to unmask how propaganda is produced in the private sector under conditions of the capitalist mode of production.

Commodification under CoP has at least two new meanings compared to Marx's notion. First, under CoP, what is being commodified is the Representation glued to a commodity, which turns out to be the selling point that propels the capitalist into the circle of CoP. In that sense, CoP is an extension of the original notion of commodification in Marxism. Second, CoP foregrounds the power relations between the state and the capitalist. This is a rather new focus of commodification in contrast to the foci in Marx's account, either the three elements of commodity, namely, exchange-value, use-value, and value, or the embodiment of human labor. Put together, CoP is meant to be a new interpretation of Marx's notion of commodification in the culture industry.

In the US context, scholars have commonly viewed commercialism as a dominant force in media production (Anderson & Strate, 2000). The marker of media commercialism is advertising, a commercialized propaganda. For example, Jhally (2000) referred to 20th-century advertising as "the most powerful and sustained system of propaganda in human history" (p. 27). In CoP, this commercial form of propaganda/publicity and state propaganda reunite in popular culture. The realm of popular culture, as noted earlier, is deeply ideological. The dominant ideology is now disguised under the pseudonym "popularity." Postman (2000) has illuminated this point, saying that:

> American television limits freedom of expression and choice because its only criterion of merit and significance is popularity. And this, in turn, means that almost anything that is difficult, or serious, or goes against the grain of popular prejudices will not be seen.
>
> (p. 51)

The framework of CoP is grounded in political economy. Within that scholarship, Political Economy of Communication (PEC) particularly deals with the power relations between capitalism and media communication (McChesney, 2013; Mosco, 2009). According to the PEC, fundamental components of a given communication system include political institutions, market structures, firms, support mechanisms, and labor practices. All of these elements are necessarily included in my discussion of CoP.

My overall analysis of CoP is dialectical. The philosophical term, "dialectical" can be ambiguous. What I mean here though is rather straightforward, as Harvey (2010a) puts it, "seeing everything in motion, in contradictions and transformations" (p. 12). The elements of PEC are to be scrutinized in following analysis not within a determined structure, but in a dynamic process leading to the production of the social space of Red Tourism. McChesney (2013) contended that "[t]he PEC is not just about making a structural analysis of communication systems and policy debates, as important as those are. Its practitioners also analyze how communication defines social existence and shapes human development" (p. 69). In light of the PEC, the latest practice in the capitalist world can be characterized by hyper-commercialism.

In his discussion of the problems in US media, McChesney (2004) maintained that an increasingly intensified commercialization of news media is shaping and colonizing the American press. He referred to this trend as "hyper-commercialism." Hyper-commercialism is driven by increasing amounts of advertising to an unforeseeably maximum extent that the traditional boundaries between commercialism and media are "dissolving" (p. 153). McChesney (2004) identified two consequences of hyper-commercialism in the United States: (1) commercialism is invading public spaces including public broadcasting, museums, colleges and universities, and the like; (2) advertising and PR are converging. McChesney (2004) noted, "in hyper-commercialism, corporate power is woven so deeply into the culture that it becomes invisible, unquestionable" (p. 167). One of the tasks in analyzing CoP is to expose such corporate power to broad daylight. Applying the concept of hyper-commercialism to highlight commercialism brought up into propaganda production, CoP, in particular, addresses the power exchange in the *inter-sector* relations between the state and the private and the interplay between capitalism and politics.

The state–capitalist bond

In the latest form of capitalism, the state behaves like a capitalist. For example, the state uses taxpayers' money to invest in infrastructures in order to generate more tax revenue (Harvey, 2010b). In examining the real estate boom in China, Glaeser et al. (2016) demonstrated that the state plays a decisive role in housing markets, if not the market itself. The state plays a similar role in developing Red Tourism and later, integrating scattered small businesses into an industry-level economy. Simply put, if the state had not used a huge chunk of tax revenues to undertake massive infrastructural projects such as high-speed railways, airports, and highways for those underdeveloped old revolutionary areas, Red Tourism could not have evolved from, say, a typical sightseeing tourist activity to sound industrial clusters accounting for 20 percent of total tourism revenues. By virtue of this, it is not an exaggeration that the state is the biggest and dominant investor in Red Tourism.

The other dominant force is capital. Harvey (2010b) reminded us that "while many agents are at work in producing and reproducing the geography of the second nature around us, the two primary systemic agents in our time are the state and capital" (p. 185). When propaganda takes the form of a commodity and when its distribution is moving from the state to the private sector via market channels, there are strong incentives for the capitalist to accelerate the speed and extend the range of such distribution. From a capitalist's point of view, this distribution process might be called circulation. The capitalist only sees it as the circulation of capital, not of propaganda. Harvey (2010b) claimed that "we have lived through an astonishing period in which politics has been depoliticised and commodified" (pp. 218–219). Harvey goes on, suggesting that "state and capital are more tightly intertwined than ever, both institutionally and personally. The ruling class, rather than the political class that acts as its surrogate, is now actually seen to rule" (p. 219). A half century ago, Baran and Sweezy (1966), too, pointed out that the power relations between the state and the capitalist were more intertwined and stronger in a monopoly capitalism system than previous forms of capitalism. Against the backdrop of militarism and imperialism (two major ways of absorbing surplus), for instance, the oligarchies tended to deceptively propagandize a very belligerent international situation in order to promote government spending on the military (Baran & Sweezy, 1966).

The state-capitalist bond can also be interpreted from the angle of postmodernism. I have found Jameson's (1991) analysis of contemporary cultural trends illuminating to reveal a globalized China, even though transplanting the idea of postmodernism from Western societies to contemporary yet traditional Chinese society is notoriously difficult. It is not so much a matter of recontextualization, which alone requires a comprehensive analysis of all possible economic, political, social, historical, and cultural conditions in China's context, as developing another postmodernism theory. In view of these difficulties, I draw only upon the part that Jameson (1991) refers to as "the cultural logic of late capitalism," leaving the rest of his argument of postmodernism open to the reader's own interpretation.

At the heart of Jameson's articulation of the dominant cultural logic is a spectrum of crises created by consumer capitalism. First and foremost, culture is in crisis. Losing its autonomy, culture no longer looks like culture but mutates into commodity via continuous commodification. The upshot is imagined as a kind of "degraded" cultural landscape, which owes a great deal to Horkheimer and Adorno's idea of the "culture industry." This leads Jameson (1991) to argue, "aesthetic production today has become integrated into commodity production generally" (p. 4). The second is a crisis of representation over historicity. "Pop history," or the mediated stereotype of the past, has gained increasing currency in "an age that has forgotten how to think historically" (Jameson, 1991, ix). To put it another way, historical authenticity is now being replaced by the simulacrum. Last but not least, the cultural issues spread out across economic and social realms, resulting in a "hyperspace," to borrow a term from Jameson (1991, p. 38). Elsewhere Jameson (2000) described it as "collapsing the cultural into the economic—and the economic into the cultural" (p. 53), and that "culture has become decidedly economic" (p. 54). It is in such hyperspace where CoP takes place. All of these crises, discontinuities, and collapses direct our attention to an unprecedentedly close bond between the state and the capitalist.

The state–capitalist partnership also manifests itself in the neoliberalism trend. Wang Hui (2003) argued that the upshot of the 1989 social movement in China was neoliberalism increasingly gaining hegemonic power. Wang (2003) goes further, characterizing China's economy after 1989 as a "dual nature": "market extremism" representing discontinuity on the one side and state guidance as continuity on the other (p. 43). This duality defines the nature of Chinese neoliberalism. Specifically, Wang (2003) regarded China's neoliberalism as a combination of elements of "market extremism," "neoconservatism," and "neo-authoritarianism," repeatedly referring to it as "the commodification of power". In analyzing China of 1997 and onward, Wang (2003) claimed:

> The links between political power and market arrangements, the recent production of poverty and inequality, and the internal links between the old networks of power and the new expansion of markets have all provided new opportunities to rethink modern Chinese history and a new creativity in understanding the discussions of the legacies of the socialist system.
>
> (pp. 97–98)

To a great extent, commodification in CoP is the commodification of the power relations between the state and the capitalist. During this process, propaganda as a social space is being commodified in conjunction with other social-economic development projects such as urban improvements, local economic development, industrial structure adjustment, and so forth. Such commodification does not happen by accident, but is a social process in the latest form of capitalist system where the boundaries between politics and economics, culture and economy, lifestyle and commodity, collapse.

Land issues

Land acquisition is of paramount importance for Red Tourism development. China's land policy is characterized by a dual land system: urban land is state-owned, whereas rural land is owned by rural collectives. According to China's Constitution (1982, rev. 2004), the *Land Administration Law* (1998, rev. 2004), and the *Property Law* (2007), farmland cannot directly enter the real estate market unless changing land ownership from the collective to the state. In other words, the state is granted the right to sell rural land to commercial developers. But there is a prerequisite: the land must be used in the public interest. In principle, acquired farmland cannot be used for any commercial purpose. In practice, the vague term "public interest" is always subject to favorable interpretations by governments at different levels. This had resulted in a "buy low, sell high" land-grabbing wave in rural China in which local governments acquire rural land at discount prices and sell it to commercial developers at a significant markup. Red Tourism, deemed as a public good significant to both public education and local economic growth, is considered a legitimate reason to acquire land.

This is nothing new. Tourism villages have been an important driving force for rural transformation development (RTD) in China, particularly in remote regions (Long et al., 2009). Tourism has brought spatial morphology evolution to Chinese rural communities. In an empirical study of a rural tourism case, Xi et al. (2015) showed that a higher degree of land-use intensity occurs in the village closer to scenic spots. Using tourism development as a pretext to grab land has been a common business strategy for Chinese real estate developers (Xu et al., 2012).

Land finance is a sign of another coming round of capitalization in rural China after township and village enterprises (TVEs) in the 1990s. Acquiring low-cost land is a way to further expand capitalism in space (Harvey, 2003b). It is considered the "most noteworthy fiscal phenomenon in China" (Wang & Ye, 2016), and has propelled the social and spatial transformation of Chinese rural communities while creating many problems (Webber, 2012).

But this fact is rarely reflected by a growing body of literature on land issues in rural China. Studies of the kind have suggested that land acquisition caused widespread social unrest in rural China (Cui et al., 2015; Li, 2016; Song et al., 2016). As to what the term "issues" denote, research in this vein focuses primarily on a series of problems, including corruption, government intransparency, undervaluation, unfair compensation, etc. Little attention has been devoted, however, to capitalism and dynamic power relations between the state and the capitalist. Song et al. (2016) identified the key stakeholders in a land acquisition case as the central government, local governments, village cadres, and local villagers, concluding that it is the village party secretary who led to villagers' tension over land acquisition; the role of capitalist was excluded from their analysis. In a similar study, Wang and Ye (2016) employed large N-datasets and confirmed that local top leaders' (e.g., the party secretary) tenure is a determinant of land finance. The latest studies of land expropriation, albeit addressing political implications, were still lacking political economic analysis. For example, Sargeson (2016) demonstrated that changing land from collective ownership to state ownership weakens democratic participation in self-government in small villages, which contradicts the classic liberal assumption of private ownership of property leading to democratization. Cui et al. (2015) suggested that land expropriation deteriorated the villagers' political trust in local governments. All of these studies left capitalists and capitalism, key stakeholders and a system, unmentioned.

The most common conclusion drawn from this constantly snowballing body of literature seems to suggest that it is the flawed dual-track land tenure system that is responsible for negative outcomes. Some suggested that if land policy is reformed, then land issues can somehow be resolved (e.g., Li, 2016). Unsurprisingly, many have called for land rights reform. But other studies have shown that land reforms in China, in spite of stimulating local economic growth, had not settled conflicts between peasants and local governments (e.g., Cheng, 2016). Capitalism, both as a mode of production and a driving force in land sales and transfers, has been systematically neglected. Such ignorance and underappreciation of capitalism in shaping social life has been pervasive in contemporary communication inquires worldwide, such as in digital communications studies (McChesney, 2013).

Among those examining Chinese land issues, Wilmsen's study is a notable exception. Wilmsen (2016) argued that what some may call the liberalization of rural land markets is "a conduit for extending capitalism in rural China" (p. 13). This argument corresponds closely to Wang Hui's (2003) critique of Chinese neoliberalism in which marketization was mistakenly treated as a potent mechanism for achieving political democracy. Due to the research's scope, however, Wilmsen does not provide evidence from the real world to illuminate this point. What's more, speaking of the capitalist mode of production without examining capitalist practices in land acquisition further weakens the argument. My discussion about land issues in relation to CoP aims at filling this research gap by focusing on the dynamics between the capitalist and the state. In particular, I look into the dual and mutually penetrated role of the state and the capitalist in commodifying rural land.

Towards a framework of commodification of propaganda (CoP)

Lately, China has witnessed the marriage of commercialism to propaganda. Red Tourism and anti-Japan World War II dramas, for example, have contributed greatly to the emergence of what some called the "Red" economy. Commercialism, the much-ballyhooed champion of capitalism and once considered the arch enemy of socialism in the Maoist era, has now become an incubator for growing a new generation of propaganda. This phenomenon is recognized but rarely realized, or seen as merely the behavior of commercialism. Understanding this unconventional marriage between capitalism and propaganda requires extensive deliberation.

Drawing from Marx's thinking of commodity fetishism and wedding that to Harvey's (2010b) notion of politics having been increasingly depoliticized and commodified, I propose an analytical framework within the tradition of the political economy of communication (PEC), which I refer to as the commodification of propaganda (CoP). CoP is an embodiment of commodity fetishism at the crossroads between propaganda, the money-making ideology of the market economy, and the capitalist mode of production.

Structural transformation of propaganda

The idea of CoP is formulated around a series of dialectical relations, including state vs. capitalist, propaganda vs. commodity, audience vs. consumers, indoctrination vs. consumption, and so forth. Altogether, CoP suggests an ongoing structural transformation in the production of propaganda: the previous relationship between propaganda and propagandee now turns into product and consumer, and therefore, passive indoctrination turns into active consumption (Figure 6.1).

A cognitive map of what I refer to as the structural transformation of CoP, the diagrammatic representation of the process of CoP entails the layering of five different levels of the structure of CoP (Figure 6.1). Note that state is not absent from the CoP framework; rather, it moves from the frontstage to the backstage. Backstage does not automatically mean unimportant; think movie director. In the case of CoP, the state still plays a pivotal role after that transformation, which I will articulate later. Structurally transforming the production of propaganda from 5 "Ps" to 5 "Cs," CoP synchronously promotes commercial values and ideological values.

It is now clear that much confusion exists on whether commodified propaganda is still propaganda after this structural transformation. First, it makes it harder for lay people, perhaps even some media practitioners and researchers, to tell whether the new product is propaganda or merely another ordinary commodity. Commenting on my research paper about China's propaganda, a reviewer of one of the major journals of the field asked me whether I would consider the "Red-themed" Chinese restaurant at Chicago's China town "propaganda." The reviewer believed that the restaurant's decoration with Mao's images all over the place is just an instance of commercialism, detached from propaganda. In my encounter with a group of Chinese television directors and practitioners, all of whom have master's degrees in film studies, they rejected the association between China's

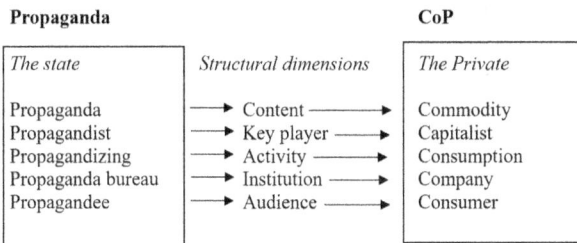

Propaganda		CoP
The state	*Structural dimensions*	*The Private*
Propaganda	→ Content →	Commodity
Propagandist	→ Key player →	Capitalist
Propagandizing	→ Activity →	Consumption
Propaganda bureau	→ Institution →	Company
Propagandee	→ Audience →	Consumer

Figure 6.1 The structural transformation of propaganda. Created by the author.

Anti-Japanese War films and television dramas with propaganda. Again, they, almost unanimously, reckoned that those shows are nothing but an instance of commercialism. That is, as often as not, how we see a commodified social space: an apolitical fantasy without thinking of the Representation, or ideology, for example, the Disney theme park.

The second problem is rather theoretical. Upon propaganda being produced as a commodity, it renders traditional content based and oriented propaganda analysis inadequate. For example, economy, labor issues, and the financial dimension are largely missing in traditional propaganda analysis. Therefore, we need a new framework. CoP is a remedy in this regard.

Rather than referring to the producer of commodified propaganda as entrepreneurs or businesspeople, I call them "capitalist" throughout this chapter for a few reasons. First, CoP requires substantial capital. In practice, this is largely surplus capital from capital-intensive industries such as real estate and the regular tourism industry. That is, the CoP itself is capital intensive. CoP is not about the innovative entrepreneurial activity through which capitalists extract profit by producing commodities with propaganda content, such as T-shirts with Mao's image or soldier's caps with red stars sold at touristic sites of the Great Wall. These are merely manifestations of commercialism and commercialization. CoP, by contrast, points to an industry that is parallel to many others under the same capitalist mode of production. Second, calling the producer "entrepreneur" might invoke a connotation of innovation, something often referred to as "entrepreneurship." This is not necessarily wrong. For one, entrepreneur is one of the roles that the capitalist plays in the CoP. For another, to successfully run the Red business, the capitalist needs to be innovative, for better or worse. Nevertheless, calling a capitalist "entrepreneur" diminishes the role of capital in this CoP setting.

Part of the social space of propaganda, CoP contains dialectic relationships across political, economic, cultural, and social realms. As such, CoP rejects any single-factor determinism (e.g., the party's ideology, commercialism, neoliberalism, and postmodernism). I do not simply call it "commercialized propaganda" for several considerations. First, commercialism can easily belie the delicate relationship between the state and the capitalist. While commercialism does, too, emphasize the power of capital, it pays little attention to state power. Consider advertising and PR, two notable markers of commercialism. They direct us to corporate/private propaganda and no other forms or sources. It is true that CoP involves commercialization insofar as propaganda content is commercialized when deployed by the capitalist, but this is only one layer of CoP, perhaps, the outermost yet most deceptive one. It evades the critical question as how propaganda becomes a commodity. Secondly, commercialism leads many to believe once commercialized, propaganda ceases or diminishes. This is the distinct impression I received from many interviewees of this study which is noted earlier. In contrast, the notion of commodification takes us to a different, pseudo-religious realm characterized by Marx's notion of "commodity fetishism." It is "religious" not only because of the fetishism of transforming social relations into commodities but also because, like other religions, it creates a hierarchical system with which capitalism continues dominating.

At the core of CoP is the nexus between state power and capital power. It means that CoP can only be understood by examining the dynamism between the two types of power. A simple condemnation of either commercialism or state propaganda is misleading at the least. As such, my examination of CoP is consistent with the notion of propaganda as a social space. CoP is not static but fluid. It is not merely about the individual role of the state, of the capitalist, of propaganda, and of the audience, but the relations between all of them. This is a vital point that cannot be overemphasized. I will revisit this at various points as my analysis of CoP proceeds and the case unfolds.

Major stakeholders

There are six major stakeholders in CoP, namely, the state, local governments, the capitalist, financial institutions, laborers, and consumers (Figure 6.2). Their network crosses two sectors: the state and the private, and at least two spheres: the economic and the ideological.

In the CoP, the state has a dual identity. First, the state by nature is the hegemon that exercises hegemonic power and is responsible for the dissemination of intended ideologies. The state also introduces regulations and provides guidelines to ensure how propaganda is commodified and then some. The second role of the

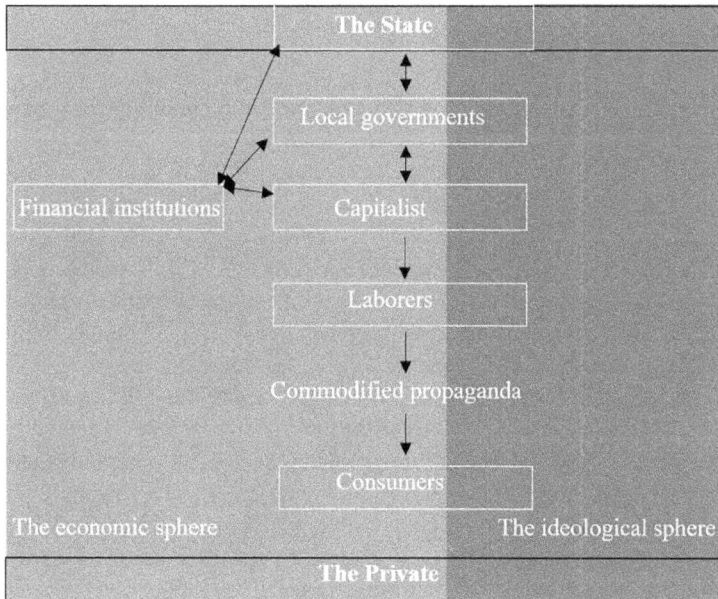

Figure 6.2 The network of stakeholders of CoP. Created by the author.

state is the investor. The state acts as the biggest investor in CoP. It is based on these roles that the state is contractually bound as the capitalist to commodify the social space of propaganda. In addition to investing in communications infrastructure, the state has absolute power in land expropriation. Harvey (2003b; 2010b) noted that spatial relations have been shaped by capitalism. What is yet to be said is the state's role in the capitalization of land. The dual-track land policy in China grants the state monopolistic power in acquiring, transferring, and selling land. Duckett (1998), writing on real estate and commerce bureau in Tianjin, China, has demonstrated that contradicting neoliberalists' expectations, the Chinese state bureaucracy can wholeheartedly embrace a program of market reform. Duckett (1998) took such enthusiasm as "state entrepreneurialism," which connotes entrepreneurial activities undertaken by the state.

In the real estate industry, the state is effectively a failure-proof investor in the wake of being insulated from any economic risk on its part, a defining characteristic of entrepreneurship. For example, the state acquires land from the rural collective at cheap state planning prices and resells it at higher market prices to commercial developers. The investor role of the state is also evident in managing public fund and state power to produce commercial propaganda (e.g., PR, ads) and regular news coverage for promoting the business, all of which ordinary capitalists who invest in Red Tourism are generally unwilling to do by themselves. Unlike in other businesses, capitalists involved in Red Tourism keep a low profile. They value their relationships with the state and local governments much more than commercial publicity.

Like the state, the capitalist's identity is also twofold: an individual serving their own interests and a proxy for the state. The dual role of the capitalist manifests a paradoxical identity epitomized by two distinct behaviors: profit-chasing and ideology-spreading. The former is always rational and primary, so much so that continued profits need to be secured; the latter, relatively irrational and secondary.

The dual role requires a dual capacity. The first is a capacity of successfully maneuvering in the capital-intensive industry sector. The second is a capacity of swiftly swimming in the waters of bureaucratic politics in the officialdom. The two capacities indicate two kinds of capital. The marker of the first capacity is investment capital, which is required in vast quantities. The second capacity is characterized by Political Capital, a special form of *guanxi* 关系. Roughly referring to social networks of the individual mobilized for doing business in China, *guanxi* now has a specific meaning. It is close to the idea of the political capital except that the influence is from a businessman not from a politician. I capitalize the term "Political Capital" to denote this particular capacity of a capitalist and to differentiate it from political capital of a politician (Figure 6.3). In the CoP, Political Capital refers to a kind of invisible currency a capitalist uses to mobilize government and public resources for their business. Tied up in mutual interests between the capitalist, the state, and governments at different levels in economic activities, Political Capital is accumulated by maintaining good relationships with the government. Li et al. (2006) argue that Chinese private entrepreneurs'

Guanxi-based Political Capital

political capital (in the general sense)

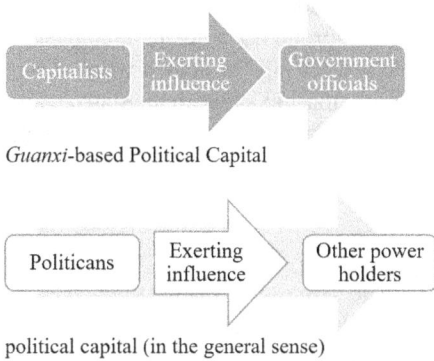

Figure 6.3 Guanxi-based Political Capital vs. political capital. Created by the author.

political participation is largely motivated by hostile institutional environments such as bureaucratic inflexibility. In examining a Red Tourism case, Zhao and Timothy (2015) viewed *guanxi* as an invisible and undocumented power that is fundamental to tourism planning and development. However, they depicted *guanxi* as social networking among different levels of governments and over-looked *guanxi* between the state and the capitalist. It gives an illusion that Red Tourism development is merely a matter of interactions among governments at different levels. What is missing from this otherwise illuminating idea is the capitalist mode of production in producing the social space of Red Tourism. This is where the CoP comes to the fore. Compared to other business sectors, Political Capital is more critical to this Red industry. I will return to this argument later; several cases will illustrate the point that the CoP represents a joint venture by the state and the capitalist.

The dual role of the capitalist has attracted much attention, contention, and confusion. On the producer end, the speculative and innovative entrepreneurial activities manifested in commodified propaganda render capital accumulation dynamic while keeping propaganda in motion. On the consumer end, revolutionary history is commercialized to the extent that commodified propaganda was as not so much to fulfill post-socialist nostalgia and stress as to profit seeking. Frivolous Anti-Japanese WWII dramas, for example, have drawn strong condemnation across both media space and social media. Some were even pulled from the airwaves over ludicrous plots and bizarre scenarios (*China Daily*, 2015). Confusion arises when people, as noted earlier, thoroughly discredit it as propaganda.

Like Janus, the state in the CoP faces two directions simultaneously: the authentic revolutionary past and a lucrative future. The contradicting interests of the state alongside the dual role of the capitalist points to a central conundrum of the CoP: a somehow "harmoniously inharmonious" relationship between the two co-ventures. They are harmonious inasmuch as the capitalist is endowed with power and resources to pursue profit, who in turn helps spread out propaganda by repacking, customizing, rebranding, and marketing it in the private sector. Inharmonious situations occur when the propaganda value in the production of commodified propaganda is depreciated or compromised by commercialism, or the other way around in which a profitable project fails to launch when it runs afoul of state regulations. This is to say, the relationship between the state and the capitalist in the CoP setting has always been contingent. In other words, internal contradictions seem to be inherent and irreconcilable in the CoP.

Preconditions and consequences

There are some preconditions for the CoP to come into being. First, there is widespread dissatisfaction with reality that leads to the emergence of a booming market with an appetite for a certain ideology (e.g., American conservatism in the era of Trump). The corollary of the CoP is that there is a market for consumption of propaganda. And the market should be large enough to attract capitalists on the one hand and to develop an industry to its full potential on the other. In light of the social inequality caused by capitalism, Red Tourism, for some, stands as a commodified social space of what might be called "Maostalgia," a nostalgia for old Maoism. Second, the CoP is capital intensive. I will elaborate this point later in my account of a Red Tourism development case. Last but not least, a "spirit." It is something like a counterpart of Max Weber's (1958) "spirit of capitalism" in *The Protestant ethic and the spirit of capitalism*, except that in the CoP such spirit is marked by the linkage between ideology and capitalism.

It is noteworthy that an authoritarian political system is not a precondition for the CoP. Although I develop the framework of CoP in China's context, it is also applicable to Western democracies. For example, the Chinese "element" in US popular television dramas and films oftentimes is politicized in such a way that the official anti-Chinese government propaganda is commodified and embodied. They are best seen through the omnipresence of vicious Chinese spies in such hit American spy shows such as *24*, *Blacklist*, *Person of Interest*, and so forth. A cultural form of state propaganda manufactured in the private sector, these shows are also instances of the CoP.

The commercialized production of propaganda shifting from the state to the private has at least two major consequences. First and foremost, propaganda after the CoP process no longer resembles its former self. Related, commodified propaganda becomes even more powerful and penetrating. I have discussed this matter earlier and assert that it is still in need of reiterations and clarifications.

The communicational territory of CoP is more of imagination than of persuasion. Commodified propaganda typically operates on collective memories. Unlike dull propaganda content produced by the state, commodified propaganda works powerfully on public imaginations of a wide range of crucial concerns including national history, national identity, cultural identity, international relations, to just name a few. By virtue of this, commodified propaganda exerts effects typically through popular media and culture. This echoes an earlier point that I have made that popular culture is a popular site for popular propaganda. It is popular culture that makes commodified propaganda unbeatably powerful. The US news media is said to help shape the American public imagination of China tremendously (MacKinnon & Friesen, 1978). US news reports claim to be "balanced" and "unbiased," though many media critics hold the opposite view. Herman and Chomsky (1988/2008) see US news media as propaganda filtering machines. Rarely do people see popular culture that way. The US audience see the pervasive and prevailing Chinese espionage scenario lurking in US popular television thrillers as an instance of commercialism at the most, while few bother to point them out as propaganda. A trivial yet equally very interesting footnote is that the presumably mainland Chinese actors in those shows always speak broken Mandarin or Cantonese, which is rather painful to mainland Chinese audiences. Except for some Chinese Americans, rarely do other American audiences demand any degree of realism for such scenes. This is reminiscent of popular Chinese Anti-Japanese War dramas in which American characters, sometimes played by international students, speak English predictably with a pronounced British or European accent. It seems that commodified propaganda only works on others. People are well equipped by their governments to detect "others'" propaganda while at the same time completely ignorant at facing "our" own propaganda. In discussing the VOA case, US media critic Jim Naureckas (2015) made a point that "they" have "propaganda," whereas the United States has "public diplomacy."

Red Tourism: an industry like no other

Many interviewed in this project have pointed out a salient fact that historic tourist sites themselves in Red Tourism are not profitable. This is even more true in popular Red Tourism provinces such as Jiangxi. Cities and provinces rich in Red Tourism sites market, brand, publicize, and promote Red Tourism, but they do not solely rely on Red Tourism sites to generate revenue as they do not charge admission fees for access. Instead, locales depend on a group of Red Tourism-oriented cottage industries, together forming the so-called "Red" industry.

It is an industry like no other. Aiming at ideological education, the state oversees the unprofitable part of the business as it protects, maintains, renovates, and manages historical sites and museums via massive subsidies. In contrast, the capitalist runs the profitable part. Some have suggested that local Red Tourism authorities can profit from regular tourism business ventures such as selling souvenirs or putting on profitable shows like the capitalist (Yi & Yu, 2013). Nonetheless, revenue from those is negligible. Profitability has always been the domain of the capitalist. Laura, a senior official of the tourism bureau of a metropolis responsible for Red Tourism, unravels the mystery. Laura explains:

> I doubt there is one [Red Tourism bureau] that would really care about earning money from Red Tourism. Sure, they can generate profits from investing in some profitable Red businesses, but what is the point? All Red Tourism bureaus are fully subsidized by the government, so they do not have any investment capital. Even though they can run the business and they run it

successfully, every penny they make must go to the government; they cannot use it. So, I just cannot see the motivation here.

(interview, December 17, 2015)

For the state, Red Tourism is not about profit. This is another salient point from the government officials interviewed during this research. Red Tourism can only merge into an industry in a handful of cities and provinces with a significant connection to the Chinese Revolution. Elsewhere, the survival of Red Tourism crucially depends on the state's support and government subsidies. Additionally, enormous investments by the state in Red Tourism infrastructure such as airports and high-speed rail in small towns and cities have exclusively focused on social gains without considering much economic gains. But there was a problem with the freebie.

Free admission did not automatically lead to an increase in spectatorship. Sometimes, the effect might be unintentionally opposite. Case in point, one revolutionary museum previously charged 40 yuan ($6.2) for a regular ticket. A tour operating company would receive a 25-yuan ($3.8) commission for every tourist they bring to the museum. In virtue of this practice, tour guides were motivated to take tourists to the museum. After the admission fees were removed, visitors declined for several reasons. From the tour company's perspective, losing commission eliminated the financial incentive. On the museum side, there was no point to draw more tourists to the site as neither 100 nor 1,000 tourists a day makes any difference in revenue. As far as day-to-day management and maintenance were concerned, the fewer tourists the better. On the tourist end, those who brought the tickets themselves are motivated to see and learn from the museum. By contrast, those coming to the museum with free admission sometimes might not be enthusiastic about the museum or the subject at all. For instance, in those scorching summer days, this museum was usually crowded with emigrant workers who came here simply to cool off. From the tour guide's point of view, Red Tourism is unprofitable too. In China, the major source of income for tour guides does not come from their services but commissions, generally from selling tickets and taking tourists shopping and/or visiting additional attractions. Since Red Tourism sites are free of admission, guides have an incentive in avoiding them.

In light of its unprofitability, some local governments have lost initial motivation for developing Red Tourism, viewing it as a pure propaganda project, no longer a promising business opportunity. Laura explains:

We came up with a Red Tourism development plan in 2013 but have not yet made it public. Basically, we designed several main routes with each featuring a corresponding theme, including "Road to Victory" and "the Anti-Japanese War." Well, those were routes, but they were very much labels indeed. Because we did not have any specific promotion plans for travel agencies, they did not pick the routes. They, at best, would choose one or two sites if those are conveniently on their own major routes. It is very true that no tour operator would consider Red Tourism sites as serious as regular touristic sites.

(interview, December 17, 2015)

Kevin, an official of a tourism bureau at the provincial level, even disputed the term of "Red Tourism industry." Kevin made a point that except for a few Red Tourism-rich cities such as Yan'an, other places have made a determined effort to develop Red Tourism only because they lack natural resources to develop regular tourism. Kevin explains:

> Although people treat Red Tourism as an industry, it only accounts for a small percentage of overall tourism revenues. It is about ideology. What we call Red Tourism is merely a form of patriotism education in other countries. For example, the Lincoln Memorial, in this regard, is a Red Tourism site because it also aims for spreading the ideology.
>
> (interview, May 30, 2014)

Characterized by a mutual overlap between private enterprises and public institutions, the Red Tourism industry is a mixture of cultural industries and public/government institutions. Nevertheless, the two domains cannot simply be divided between the private and the public as they appear. Under the current situation, cultural industries in China, though by and large private, are always entangled with public/government institutions. Kevin remarked that every time he attended the national conference on Red Tourism, he found most attendees were those who were from the cultural sector in charge of venues and museums. They are part of cultural industries and part of public institutions.

Most venues and museums are subordinate *shiye* units of the Bureau of Cultural Affairs (*wenhua ju* 文化局) and other government departments. For example, the Bureau of Cultural Affairs runs museums, and the Bureau of Civil Affairs (*minzheng ju* 民政局) is in charge of revolutionary martyrs' cemeteries. When the government manages Red Tourism resources, it always involves both industries and units within public institutions. The Tourism Bureau does not have any tourism resource under its name. What the Bureau does is merely to provide platforms and establish some industry standards. It has no control of any tourism business such as operating a tourist site or an attraction due to lacking of administrative power. If a privately owned enterprise invests in a Red Tourism spot, what the Bureau can offer is to help the company build a few public facilities such as public toilets. The Bureau does not have the budget specific for Red Tourism. But the department of propaganda and the Bureau of Cultural Affairs do.

I now turn to my CoP cases. What follows is not to be read as a description of day-to-day operations of the Red Tourism business. Rather, I intend them as "documentaries" to showcase individual events and issues pertaining to commodifying the social space of Red Tourism, through which the conception of the CoP comes to the fore.

The reenactment show: revitalizing capitalism

Henry, an entrepreneur of the Red Tourism industry, has invested in multiple on-site reenactments at scenic Red Tourism spots nationwide and monopolized the market. The Red Tourism market had grown large in 2005 when Henry invested RMB 30 million ($4.62 million) of capital accumulated from years of his ordinary tourism business in the venture. Realizing that the majority of Red tourists were mainly post-'80s and post-'90s youth, he aimed to depart from old-style Red Tourism's dull explanations and dry exhibitions and instead adopt the interactive trend. The success of this venture brought Henry acclaim in his business circle.

While all investments are risky to some degree, private investment in Red Tourism is much riskier. Henry explained that for investors in nature-based tourism, it is rather a matter of tourism management upon the completion of the investment. Red Tourism investment, by contrast, requires a considerable amount of manpower, resulting in higher management costs and higher production costs. Natural attractions also publicize themselves over time, which is not the case for Red Tourism. A contract for running a regular tourist business in natural attractions will nearly guarantee a profit, but a contract in Red Tourism will not do the same. Once Red Tourism construction work is completed, substantial following investment for generating publicity is still needed. According to Henry, the risk of investing in Red Tourism is several times higher than that of investing in regular tourism. Henry says:

> Commercial propaganda (publicity/ads) is pretty much on our own. The government only promotes propaganda on the macro level, say, the development of the Old Region or the sacred place of Yan'an, and celebration of Mao Zedong's thirteen years in Yan'an. But our investment is on the micro level, which the government will not propagandize.
>
> (interview, June 4, 2015)

Besides, Henry suggested that the market also completely depended on his own ability to generate demand. It requires a certain kind of maneuver and strategy, including a capacity in taking advantage of government politics.

Henry's war reenactment business changed its themes according to national politics. In light of the Diaoyu islands dispute between China and Japan coincident with an easing of military tension between mainland China and Taiwan, the mainstream Civil War-theme television dramas were replaced by Anti-Japanese War ones. Correspondingly, the previous hostility between the CPC and the KMT was revised to reconstruct a kind of partnership between the two parties in fighting against the Japanese troops. In other forms of cultural industries, for example, television and film, such partnership is now typically portrayed as some kind of romantic relationship between two spies from the two different political camps. Henry explains:

> My current strategy is anti-Japanese stuff. I have my own thinking: first, I saw the relationship between the KMT and the CPC was bettered; second,

the Japanese along with the Americans now keep troubling China and what the Chinese people hate most is the Japanese. So, I believe switching to this (anti-Japanese theme) is easier for people to accept.

(interview, June 4, 2015)

Henry had two projects underway when I interviewed him in summer 2015. One reenactment site in Chongqing, which features "the Sichuan Army leaving Sichuan to join in the Anti-Japanese War," was near completion with an investment of RMB 50 million ($7.7 million). The other, a more ambitious reenactment, the Battle of Tai'erzhuang, just passed project evaluation and was in negotiation over development details with the local government. It should be noted that unlike other Red Tourist attractions, the two projects share a marked similarity that both represent spirited resistance of the armies of the Republic of China against the Japanese armies, not the Communist-led troops. The Sichuan Army was a local military force relatively independent from the central government, whereas the KMT troops played the leading role in the Battle of Tai'erzhuang. In other words, the two projects have less to do with the CPC and are considered "Red" only because of their patriotic theme, an extension of Red Tourism. Rather than representing the CPC legacy, the two reenactment sites aimed to propagate a positive national image or some positive collective Chinese identity.

The risk, however, is by no means merely metaphorical. The production nature of reenactment shows in the Red Tourism sphere poses constant hazards to workers. Companies promised that the work is absolutely safe, stressing that prop guns used in the reenactment are loaded with blank ammunition that contains gunpowder but no bullet or shot. Nevertheless, since there were a lot of reenactors who performed daily, injuries such as burns, abrasion, dislocation, fracture are almost inevitable. Only when the injury is serious enough to prevent the injured from working can the worker take a paid sick day; they have to continue to work otherwise or they can take a day off without pay.

Raw materials in the reenactment business simply mean gunpowder and explosives. To reenact an authentic version of revolutionary battles, the crew of the show actually used more than two hundred genuine rifles, six cannons along with armored vehicles that were once used by both the KMT and the Communist forces during that time. To achieve greater visual impact, each hourlong performance used 800 blanks and 80 shells (Liu & Chen, 2012). Since the show performed each day during the tourist season (May to November), the demand for gunpowder and explosives to be used in the production is large and crucial. Given the grim situation of counterterrorism and national security worldwide, transporting and storing such large quantities of ammunition alone requires extraordinary Political Capital to appease the government, the military, and the police, not to mention obtaining a requisition for explosives and bullets. Rarely can any group of entrepreneurs or private companies get this kind of support from the state. This again shows the delicate relationship between the capitalist and the state in the CoP.

Tragedy strikes when things go even slightly wrong. In 2011, the entrepreneur who invested in and ran the *Yan'an Defense* reenactment show was commended as the "first person of China's Red Tourism" in the National Red Tourism Conference for the success of his reenactment business in Yan'an. The following year, he invested RMB 100 million ($15.4 million) for a similar historical reenactment named the "Liaoshen Campaign" in Fushun, a northeastern city. The reenactment was of the Liaoshen Campaign, the first of three decisive campaigns during the War of Liberation took place in 1948 and signaled the final stage of the PLA's offensives against the KMT. To a great extent, the Liaoshen Campaign reenactment is a replica of the *Yan'an Defense*: the same promised "real guns and live explosions," homogenous scenes, and recruitment of a famous director, Zhang Qian who directed the well-known Anti-Japanese War television drama series *Liang Jian* 亮剑 (Unsheathing the sword). This time, the entrepreneur would face considerable adversity.

Seven people were killed in an explosion after a dress rehearsal on April 28, 2014, a few days prior to the opening of the reenactment show. The explosion was set off by two pyrotechnicians who were handing explosives for special effects. The two pyrotechnicians died on the spot alongside three staff members of a local administrative committee who were passing by and two workers (China Radio International, 2014).

Two months after the fatal tragedy, a follow-up news report in a local newspaper titled "An accident makes the Liaoshen Campaign reenactment more riveting" (Yici, 2014) made a "soft sell" of the performance. It should be noted that commonly dubbed as "soft news" (*ruanwen* 软文) in the lexicon of Chinese media professionals, these paid articles have been a thorny issue in the Chinese news industry (Tsetsura, 2015). In 1997, the department of propaganda along with other government agencies launched a long-lasting anti pay-to-publish campaign, requiring that news must be distinguished from advertisements and all forms of paid news are forbidden (Central Propaganda Department, 1997). However, the situation did not improve until violators were met with heavy punishment, ranging from disbarment to imprisonment. Today blatant paid news in Chinese newspapers is very rare. Li and Li (2014) referred to this rarity as a manifestation of "soft power," pointing to the capacity of people in power to dictate media practice. One year after the death of the seven people, the Liaoshen Campaign reenactment show celebrated its opening with more than 3,000 tourists visiting the site on April 23, 2015. Brimming with photos and text, the local newspaper covered the opening event with a full-page display (Zan, 2015).

Ostensibly a safety issue, the accident also points to serious labor issues in the CoP. Mike was having a difficult time recruiting laborers (reenactors) when he started his Red business. Reenactments by their nature are a form manual labor that requires actors to work exposed to the elements. Like in other labor-intensive industries, capitalists aim for cheap labor power in the CoP, typically, landless farmers. Mike says:

> Wherever we go, we are welcomed by local governments because most people we hired were locals, those uneducated [farmers]. As long as they can run, we employ them. They cannot be hired by any other work units because

they don't look good and are poorly educated. What they have is labor. One can only make money from trading his labor, if you will. They don't have good physical strength and what they can do is mere running errands. So, this [business] has solved the unemployment of these people.

(interview, June 4, 2015)

Considering that the reenactment work is physically demanding, yet underpaid compared to jobs, local young male workforce would prefer to go out to work in cities. As a result, women and the elderly formed the bulk of workforce in the Red businesses. This also accounts for the abundance of female soldiers in the battle reenactment shows.

The *Yan'an Defense* appeared in the BBC 2013 documentary, *China on Four Wheels*, which helped the scripted reenactment gain international popularity; it did not help the show extend market reach though. In 2011, the show performed three times during the day and once at night. Starting from 2014, the reenactment only performed once a day. Previously there were several horses for the reenactment show, as shown on the billboard, later only one horse and three donkeys were left. With a decrease in visitors, the company had increased the ticket price from 150 yuan ($23) to 180 yuan ($28).

Workers' payments have been rock steady during the last decade since the opening of the site. The monthly wages of regular reenactors were 1500 yuan ($231) and 1800 yuan ($277) for drummers in 2016. All workers were paid without the compulsory social insurance by law. The wage rates were so low that they could barely make ends meet. Note that in 2015, the per capita disposable income of urban residents in Yan'an was RMB 33,127 ($5096) (Statistical Bureau of Yan'an City, 2016). Also note that the cost of living in Yan'an is much higher than a similar city because of the tourist flats. For example, a bowl of ramen cost about $3 in Yan'an, whereas it was only $1 in Baoji, the second largest city of Shaanxi Province.

There were no holidays, weekends, and paid sick days for all workers, reenactors, and other laborers. All were expected to work every single day during the tourist season (from February to November) regardless of weather conditions and their physical conditions. Previously a drummer in the reenactment show, Carter is now a security guard working in the site. According to Carter, the show performed each day during the tourist season without a single day break in the last 11 years even under adverse weather conditions, and even if there were only about one or two hundred spectators. Every single day in the nine-month tourist season was a full workday for the workers. Sometimes reenactors had to perform in heavy rain without any additional compensation. Workers would receive a monthly 200-yuan ($31) bonus providing that they do not take sick leave in that month. If they do, they will not get paid for that day plus losing the bonus. Taking that into account, a sick day will cost a worker 250 yuan ($38), or about 1/6 of their monthly wage. Therefore, it was extremely rare for a worker to take sick leave even when they were severely ill. Following a common practice from China's manufacturing industry, the reenactment company provided workers with meals and lodging on the reenactment site. Not many workers chose to live in the dormitory; the majority of workers are locals, who unlike migrant workers have family and relatives nearby.

The workforce of the reenactment company is roughly divided into two groups: ordinary workers and support staff. Roughly speaking, workers did manual labor, whereas staff members were in charge of business management and other admin-istrative work. The wage gap between the two groups was huge. Staff members earned about four times as much as regular workers, plus bonuses. Since the site sat empty during the winter months, workers left the site over winter for other work. Unlike full-time staff, Carter relocated his family to Yan'an city to pursue a better education for his son. They lived several years in a rental house, which left them little money after rent and food, let alone skyrocketing education costs in urban Chinese schools.

Like Carter, most of the reenactment show's workforce were locals or from nearby counties. Recruitment of reenactors was not challenging. The company stopped recruitment after quickly attracting 300. According to Carter, had recruit-ment continued, even more people would come. This is because women from villages have limited employment prospects. Reenactors need to work 8-hour a day like elsewhere even though the show only lasts about one hour. Every day workers reported to work in the morning, preparing and then performing the reenactment show. After having their lunch in a makeshift canteen in the loess hill, workers came back to the site to take a roll call and then killed a few hours before the end of the day's work. In addition to regular performance, rehearsals, though infrequent, took place when the producer saw issues in their performance. Reenactment is far from the only profit center of Red business. In fact, real estate is much more lucrative, a telling fact about the CoP.

Golden Yan'an

Land development is one of driving forces behind Red Tourism investment for capitalists. Huang (2008) has noted that real estate is the "most political sector in the Chinese economy" (p. 229). It is also the site where the development of Red Tourism thrives.

The unwritten golden rule for private investment in Red Tourism is that the capitalist must ensure, prior to signing a contract with the government, a double-size-investment return from the government. For example, if one invests 3 billion yuan in a Red Tourism project, he has to secure 6 billion in return. The 6-billion return may come in many forms, for instance, building supporting infrastructure such as highways, airports, speed-rails, and other mass transportation systems in conjunction with the project. Among many kinds of rewards, land acquisi-tion is a prominent one for the capitalist to partake in Red Tourism development projects such as building a touristic site. Given the skyrocketing housing prices and increasing difficulty of acquiring commercial land nationwide, the capitalist can lock in substantial profits from land finance even before the project starts. Typically, they will allocate a portion of acquired land from the government, often by fiat, for the Red Tourism site development, leaving the rest for other commercial developments. In some cases, the capitalist would simply resell the land to other real estate developers.

Henry is not a real estate developer, but he has friends in that circle. A Hangzhou-based real estate guru recommended Henry to invest in a Red Tourism development project called the "Old Yan'an," a modern replica of the old Yan'an City in the revolutionary period. He told Henry that it was much better to invest in the Old Yan'an project than a regular real estate one because it is almost certain to kill two birds with one stone: once the project was completed, shops in the Old Yan'an will be rented out and tickets can be sold to tourists at the entrance gate. Thanks to the two separate revenue streams from real estate and tourism, Henry invested into the project. Four years after signing the contract, Henry's company had completed the conversion of agricultural land to commercial land. However, the construction site was empty when I took the interview with Henry in 2015 due to financial problems. A year after that, the Old Yan'an project changed hands from Henry to a corporate group. The name of the project also changed to "Golden Yan'an." These changes point to a battle on the financial capital front and another one on the Political Capital front.

Conveniently located at the northern extension of the main Red Tourism route of Yan'an city, Jingjiawan Village enjoyed a reputation for rural tourism, which helped the village being included in the list of "Shaanxi Rural Tourism Demonstration Villages" in 2010 (CNTV, 2011). In particular, Jingjiawan was famous for *nongjia le* 农家乐 tourism, literally, "joy of farmers' families." It is a form of farm-household-based agritourism emerging in post-Reform China. A bottom-up type of tourism, *nongjia le* offers urbanite tourists nostalgia tourism with an experience featuring rustic farm-made food and countryside-style lodging alongside other farming-related activities and provides local farmers with an additional, valuable source of income. Moreover, it also stands as a communication event insofar as social boundaries between the urban and the rural are constantly being reconfigured and negotiated in the tourist encounters (Park, 2014). About one in three farm households in Jingjiawan participated in the *nongjia le* tourism business (Tian, 2014).

This however is no longer the case. Due to a huge development called the "Yan'an Sacred Land Valley," the villagers lost their land for doing their rural tourism business. *Nongjia le* tourism that supports the needs of the rural households was seen as "primitive" and replaced by the large-scale tourism investment featuring revolutionary heritage. It is for this reason that the Jingjiawan villagers organized a dozen protests against the construction project. Their attempts, however, failed on April 23, 2014, when construction work of the project resumed after having stopped for a year due to the protests. A commentator predicts, "in the near future, this tourism village, which is well-known for flower farming and *nongjia le*, will completely cease to exist; the replaced will be a large-scale Red Tourism project with an investment of ten billion yuan" (Tian, 2014).

The company responsible for the development is Shaanxi Tourism Group (hereafter, STG), which monopolizes the Yan'an Red Tourism market from live performances to other tourism cottage businesses. Established in 1998, the Xi'an-based corporate group covers four industrial clusters: tourism services, tourism culture, tourism and cultural center area development, and tourism real estate. According to the company's website, STG has more than 40 subsidiaries, 6,000 employees with net assets of 10 billion yuan ($1.54 billion), and comprehensive operating income of 18 billion yuan ($2.77 billion) by 2016 (STG, n.d.).

The Valley project was highly controversial over its real estate-centric development. Land transfers and real estate sales are cardinal components of its business model. The 10-billion-yuan project includes five major subprojects, namely, "Urban Life," "Golden Yan'an," "Northern Shaanxi Folk Park," "Cradle Valley," and "Long March Children's Theme Park."

Commentators and industry insiders believed developing large Red Tourism projects in Yan'an was not purely driven by market demand. They pointed out that Yan'an already had adequate, fully developed Red Tourism sites that attract millions of domestic tourists each year. Compared with those sites, the Valley project without much tourism appeal does not have any advantage from a tourist perspective. The ultimate driving force for the project was believed to be land finance. By virtue of this, commentators referred to the Valley project as a "land game" (Tian, 2014).

Golden Yan'an is the core among all the five subprojects. It is an ambitious project to build a new "Golden" city within a city, which has stirred up controversy. STG describes the Golden Yan'an as "the reproduction and beyond of the old Yan'an, the visible glory and dream of Yan'an, a three-dimensional city, a smart city, an eco-friendly city, and a fortune city" in the project's publicity (Sacred Land Valley, 2014). Perhaps more importantly, the craze for building new iconic architecture worldwide indicates a fiercer round of capitalist competition, as Harvey (2010a) points out:

> The selling and branding of place, and the burnishing of the image of a place, becomes integral to how capitalist competition works … Bring a signature architect to town and create something like Frank Gehry's Guggenheim Museum in Bilbao. This helps put that city on the map of attractors for mobile capital.
>
> (p. 203)

Apart from its publicity, the "true colors" of the ostensible "Red" project are also easily detectable on the panoramic map of the project for promoting sales. Except for one revolutionary history museum, the properties on the map ready for sale and transfer are commercial real estate, including eight themed-hotels (3 or 4 star), residential buildings, commercial streets, bars, a plaza, a shopping center, and business clubs.

The progress of this multi-billion-dollar Red Tourism project was rather unsatisfactory. It stagnated shortly after the project set up for a variety of reasons. The first was land acquisition. In spite of the extreme land scarcity in Yan'an, the project requires 14 square kilometers (3459 acres) of land. Land acquisition took place in

Jingjiawan Village in December 2013 with all 897 mu (145 acres) of land being converted from farmland to construction land in the name of tourism development, leaving all the villagers landless. Each farm household would receive a land compensation of 147,500 yuan ($22,692) per mu. The compensation rate fell short of the villagers' expectations. On average, every mu of land in Jingjiawan could generate the villagers 60,000–70,000 yuan ($9231–10,769) from produce and rural tourism. The land compensation is less than three years of revenue of the land itself. Note that the market price for construction land in Yan'an was about 2.5 million yuan ($384,615) per mu at the time, 17 times the land compensation rate. According to STG estimated land revenue, every mu of cleared land would bring the corresponding profit of 1.452 million yuan ($223,385) (Tian, 2014). If we multiply this by all acquired land from Jingjiawan Village, which is 897 mu, the total profit STG will be making from land acquisition alone in a single village would be 1.3 billion yuan ($200 million). Therefore, land acquisition is the perfect strategy for STG to generate guaranteed profit. But for Jingjiawan villagers, landholding provides an important safety net for them. This had given rise to local farmers' stiff resistance to land acquisition shortly after they were notified of the land compensation plan.

Apart from villagers' resistance, the local government along with its departments and agencies showed marked reluctance to cooperate with STG in land acquisition. The motive of the Valley project was openly questioned, mainly by the Pagoda District Government and Yan'an Land and Resources Bureau. The local officials insisted that STG aimed at land finance under the guise of Red Tourism development. A fiduciary arrangement made in 2014 between STG and Sino-Australian International Trust on the Valley project shows that the estimated revenue from Phase one of the project is 7.939 billion yuan ($1.22 billion), which includes the land transfer income of 5.559 billion yuan ($0.855 billion), tourist ticket sales of 870 million yuan ($133.8 million), and rental income of 1.51 billion yuan ($232.3 million) (Huaao, 2014). These numbers suggest that 89 percent of the revenue comes from land finance. An industry insider revealed that the Golden Yan'an project, the major subproject of the Valley project, would lead to a land value uplift of the proximity, which in turn, would help STG reap greater profit in its later land sales and transfers.

Nevertheless, the local officials' resentment towards the Valley project did not lead them to oppose land finance in general. Both the local government and the Land Bureau maintained that they should be the beneficiaries of land finance, not STG. For one, the rural land policy grants the local government discretionary power in the chain of land acquisition, conversion, and land sales. For another, land sales have been the most important source for local government funding. While in principle the beneficiaries of land transfers and land associated with the Valley project should be the Pagoda District Government and the Land Bureau, most profits earned from land finance however went to STG in accordance with the company's business model. This was the reason behind local officials' lukewarm attitudes towards the Red Tourism project.

The third problem was financing, which was as serious as resistance from the villagers and the local officials. One local official pointed out that STG did not have sufficient financial strength for this 10-billion-yuan project. According to information released from Sino-Australian International Trust, STG reported total assets of 3.01 billion yuan, net assets of 650 million yuan, and a debt ratio of 78.4 percent in 2012 when the company obtained the Valley project (Huaao, 2014). It is no wonder that STG sought financing to build the Golden Yan'an on the slogan "Huge Financing, Financing for the Huge" (Sacred Land Valley, 2016). Also note that STG obtained land without going through an open land auction, a common practice in land acquisition in China. In other words, STG might not have obtained the land-use rights or would have paid more if it had gone through the proper channels. It is obvious that STG received preferential treatment and enforcement.

Contrary to the attitude of the lower (district) government, Yan'an Municipal Government showed extraordinary enthusiasm for the biggest-ever single investment in Yan'an tourism. The municipal government convened many meetings, calling for full support by all local government departments and agencies involved in the Valley project, asking them to sacrifice local interests in land acquisition and displacement and aim for "big." "Big" has a special meaning, connoting that political interests of the state must be prioritized. In 2012, a few months after the signing of the contract between the Yan'an Municipal Government and STG, the Yan'an Sacred Land Valley Cultural Tourism Industrial Park Management Committee was created in order to further facilitate and promote the project. In spite of functioning as a government agency, the Committee is surprisingly led by STG, which, again, was resisted by local officials.

STG, however, had the full backing of the propaganda department alongside other departments and agencies. The Valley project had its own website, mainly for PR and marketing purposes. In the "news" section, one piece was about an official of the department of propaganda visiting the site of the Valley project and attending the first performance of a Red Tourism show produced by STG (Sacred Land Valley, 2016). Representing the Yan'an Period, the "Red" show "Yan'an Memories" was said to be directed by a German team which employed many cutting-edge sound and light technologies. The website also highlighted visits by other officials. What's more, the Valley project was included in the list of "National Fine Selected Tourism Development Projects 2014" issued by CNTA (CNTA, 2014). In December 2016, STG Yan'an subsidiary company was awarded "Advanced Collective of National Tourism System," an honorary title bestowed on selected tourism-related organizations every five years by CNTA and other state departments (CNTA, 2016).

Despite the contention surrounding the project, Golden Yan'an has already become reality on my last visit in 2021. Surrounded by mountains and villages, the newly built tourist town operated as a new "fantasy" of the city of Yan'an, in which it melted sundry touristic elements into a single project, including history of Northern Song dynasty, revolutionary history of Yan'an, folk performances, antique shops and restaurants, elegant guesthouses, and so forth. The town sits empty during the daytime when tourists visit other Red Tourism sites and populate in the evening when tour buses bring in waves of tourists, large enough to fill a town. It is no wonder that Golden Yan'an has branded itself as the place where "China's largest courtyard-style B & B cluster" is located (Guo, 2019).

The "Golden" in its name speaks it all. As a major project of the Sacred Land Valley Cultural Tourism Industrial Park, Golden Yan'an occupies an area of 442 acres, or about half the area of Central Park in Manhattan. Guided by the planning and design concept of "mirroring history and capturing the soul of Yan'an" with a theme of "memory of Yan'an," Golden Yan'an features newly rebuilt iconic buildings of Yan'an at two specific historical periods, namely, the old Yan'an trading streets in the 1930s and shops and booths in the Northern Song dynasty more than a thousand years ago (Shengdi, 2021). It encompasses the representational space of the revolutionary history and that of a much longer history of the city, thusly entailing both Red Tourism and regular tourism.

I toured Golden Yan'an on a weekday afternoon by bus and was the only individual tourist who visited the town during daytime. After entering Golden Yan'an from the West Street, I felt unimpressed. This is partly because the town was empty and partly because the main street from west to east features Red-themed sculptures identical to those scattered throughout Yan'an. The only people I saw on the street were young trainees from a performing arts training center nearby. According to its advertisement, the center specifically prepares performers for local Red shows.

Note that Golden Yan'an is not merely a tourist attraction; it has formed eight major "strategic" business sectors, including (1) courtyard B & B business, (2) Red training base, (3) cultural and creative industry incubation base, (4) team-building camp for Chinese brands, (5) research and learning base for the youth, (6) performing arts center, (7) education center, and (8) sports and fitness center. These diverse ventures enable Golden Yan'an to market itself as "a reception hall and a new tourism business card of the city of Yan'an and the center of cultural and tourism industries" (Shengdi, 2021).

After arriving at the Bell Tower, my disappointment disappeared. Some people in historic costumes were gathering around. The structure itself is the center of Golden Yan'an where four main streets, namely, the East, West, North, and South Streets intersect. While East and West were unremarkable in both architecture and layout, North and South Streets proved to be highly compelling for visitors.

The North Street is a simulacrum of a trading street of Yanzhou (ancient name for Yan'an) during the Northern Song dynasty. Full of antique restaurants and shops, however, the North Street features three iconic buildings in the Northern Song. The first building is Jingluefu 经略府, the residence of the local military officials, which serves as a façade for a B & B guesthouse. The second is a replica of Chenghuang Temple Stage (城隍庙戏台), public place for folk artists to perform in Northern Song. As it was more than a thousand years ago, the stage is still producing various entertainment shows for tourists such as acrobatics, stand-up comedy, storytelling, cross talk, singing and dancing, shadow play, traditional Chinese magic, etc. Tourists can participate in the performance in the interactive section. The third is Kaigelou 凯歌楼 (triumphal tower), an entirely wooden building that adopts the traditional mortise and tenon carpentry and uses 520 cubic meters of wood. It is said that after each border war, triumphant generals held victory greetings, celebrations, prisoner offerings, and other activities here in Kaigelou. That is how the name "triumphal" came to be. There was an eye-catching reenactment in Kaigelou on the day of my visit. The show features Chinese traditional marriage customs with the highlights of throwing an embroidered ball from the top of the building to the tourists and a wedding parade after that (see Figure 6.4). According to this marriage custom, the gentleman who catches the ball will become the groom of the lady who threw it. The people wearing ancient clothes I saw in the Bell Tower earlier are also performers of this display.

The South Street is a replica of Yan'an during the revolutionary period. Based on the Yan'an street view in the 1930s and drawn upon historical records and images, the South Street of Golden Yan'an features a series of newly rebuilt iconic buildings including Anlanmen 安澜门 (the south gate of old Yan'an city), Xinhua Bookstore, and a Catholic church, inviting tourists to an imagined community in the revolutionary period. Xinhua Bookstore is now a state-owned book chain, which is under the authority of the propaganda department. In April 1937, China's first Xinhua bookstore was established in Yan'an. It is reported that the replica of Xinhua Bookstore in Golden Yan'an used original bricks from the historical Xinhua Bookstore in Yan'an for additional historical authenticity (Xu, 2021).

The Catholic church is another landmark of Golden Yan'an. The current building was built based on the photos taken by the Kuomintang delegation to Yan'an in 1937. There is a popular anecdote about the structure. Rumor has it that the original construction featured bizarre proportions due to miscommunication. The church was designed by foreigners and Chinese builders misunderstood the perspective in the design drawings. As a result, the Bell Tower, supposedly in the middle of the building, was built in front of it instead. The error was too late to rectify once found (Xu, 2021). The church turns out to be the only one of its kind

Figure 6.4 A wedding custom performance at Kaigelou. Photograph by the author.

Figure 6.5 The Catholic church in Golden Yan'an. Photograph by the author.

in the world. Considering its eccentricity, the church was dubbed as a "beautiful mistake" made by the integration of the oriental and the occidental cultures. In Golden Yan'an, the Catholic church is now standing as a bookstore, regardless of its Catholic appearance (Figure 6.5).

On a typical sunset, tourist groups brought by tour buses swarm in Golden Yan'an from the south entrance. Their first task is to check in at stylish B & B guesthouses arranged by tour organizers. Substantial revenue of Golden Yan'an is derived from lodging. Arriving late means these tourists must spend the night. As soon as they return to the street, the tourists turn a static place into a highly productive space. Awaiting them is local cuisine, creatively produced souvenirs, live performances, and parades featuring period costumes. In fact, Red Tourism is merely one of the many appealing options. Crowds frequent the bustling night market in the Northern Song dynasty on the North Street in as high numbers as those staying put on the Red-themed South Street. Regardless, tourists' consumption will eventually yield profit for the proprietors of the installments.

Conclusion

Propaganda has long been a commodity, in company with the commodification of social life. Rarely is propaganda thought of this way though. This is because how we see propaganda has been largely framed by state propaganda. In my lived experiences, such determining force does not come in the form of propaganda posters or op-eds of party organs or dull radio programs, but rather from theaters, movies, songs, televisions, video games, websites, social media posts, and in this case, tourist sites.

The commodification of propaganda is nothing new to the profit-seeking foundations of capitalism. What is new to the CoP, however, is how it swaps the role between the state and the capitalist. The state invests the most, including infrastructure, historical sites, and land, whereas the capitalist is responsible for producing propaganda. But the exchange creates some inevitable friction between the two.

Again, the CoP is by no means unique to China. In other parts of the world, dominant ideologies, whether classic liberalism, neoliberalism, or conservatism, cohabit with propaganda in the realm of popular culture. Tourism is no exception. The CoP offers a framework that will help the reader rethink tourism as not merely limited to Red Tourism but tourism in general, in the context of disseminating Representation or propaganda.

In a number of ways, this book aims to exceed the scope of existing work in propaganda studies by illustrating the following: that propaganda is a social space; that there is a propaganda culture in China; that tourism is a popular site of propaganda; that the capitalist can profit off popular propaganda by commodifying and selling them in the market; that although ostensibly commercial and non-ideological, commodified propaganda can be even more ideologically powerful, hence, more propagandistic; that the partaking of surplus capital in the commodification of propaganda further complicates the social space of propaganda; and that capitalism is, too, a driving force for the production of propaganda.

These are bold claims. Each one of these merits their own book, or two. The nature of the current study, however, is merely a sketch in this regard and makes no pretense to be exhaustive. The purpose of this book is to begin the process of systematically analyzing propaganda not on the basis of a certain set of ideologies or practices, but on the basis of the production of social space, addressing communication networks, political pathways, the political economy, and the interplay among them.

My approach was to undertake a study of Red Tourism, to investigate into its produced yet productive social space, as a way of redirecting our attention regarding propaganda from the state to the private, from the news media to popular culture, and from state politics to the dynamisms between that and capitalism. This goal cannot be achieved unless propaganda no longer appears as merely the manipulation of public opinion by the state. By virtue of this, I take as my starting point a Chinese history of *xuanchuan* that is broadly comparable with the early history of propaganda in the West. In their early stages, propaganda, whether in the form of Confucian preaching or of spreading God's words by missionaries, was aimed at a kind of "spiritual education", secular or religious. While this inner core of the *xuanchuan* culture survived throughout imperial, republican, socialist, and post-socialist China and remains effective today, propaganda in the Western world changed its neutral connotations, particularly after two world wars with a fixed evil image cemented in the minds of late modern Westerners. I conclude at a point when propaganda has been increasingly commodified through a particular form of joint venture between the state and the capitalist, which I refer to as the commodification of propaganda (CoP). CoP suggests an ongoing structural transformation of propaganda and a cross-boundary arena of propaganda where culture, history, state politics, and economy intersect. All of these problematize early models and theories of propaganda, suggesting a need to extend the paradigm of propaganda studies beyond the confines of a phenomenology of psychology and the interpretation of the province of authoritarian states.

References

Andersen, R., & Strate, L. (Eds.). (2000). *Critical studies in media commercialism.* Oxford University Press.

Baran, P. A., & Sweezy, P. M. (1966). *Monopoly capital: An essay on the American economic and social order.* New York University Press.

Central Propaganda Department. (1997). *Guanyu jinzhi youchang xinwen de ruogan guiding* [Regulations on prohibition of paid news]. http://www.zgqybxh.org.cn/index/article/read/aid/249.html

Cheng, H. (2016). Land reforms and the conflicts over the use of land: Implication for the vulnerability of peasants in rural China. *Journal of Asian and African Studies,* 1–15. http://doi.org/10.1177/0021909616654507

China Daily. (2015, August 20). *Anti-Japanese war dramas pulled from TV due to ludicrous plots.* Retrieved November 8, 2021 from http://www.chinadaily.com.cn/opinion/2015-08/20/content_21656428.htm

China Radio International. (2014, April 29). *Liaoshen zhanyi jingqu fasheng baozha qiren siwang xianchang fengsuo* [Seven killed in an explosion in Liaoshen Campaign scenic spot]. Retrieved November 8, 2021 from http://gb.cri.cn/42071/2014/04/29/782s4521854.htm

CNTA. (2014). *Guanyu yinfa 2014 quanguo youxuan lvyou xiangmu minglu de tongzhi* [A notice of publishing the list of 2014 national fine selected tourism development projects]. https://fashion.ifeng.com/travel/news/detail_2014_10/17/39106663_0.shtml

CNTA. (2016 December 12). *Quanguo lvyou xitong xianjin jiti laodong mofan he xianjing gongzuozhe mingdan* [The list of advanced collectives and individuals, and model workers of the national tourism system]. http://www.cnta.gov.cn/zwgk/tzggnew/201612/t20161212_809057.shtml

CNTV (China Network Television). (2011, February 25). Tongcun gonglu chengjiu liao "Shaanxi xiangcun lvyou shifan cun" [*Village roads lead to "Shaanxi Rural Tourism Demonstration Villages"*] Retrieved from November 8, 2021 from http://sannong.cntv.cn/special/lhsn/20110225/114809.shtml

Cui, E., Tao, R., Warner, T. J., & Yang, D. L. (2015). How do land takings affect political trust in rural China? *Political Studies,* 63(1), 91–109.

Duckett, J. (1998). *The entrepreneurial state in China: Real estate and commerce departments in reform era Tianjin.* Routledge.

Glaeser, E., Huang, W., Ma, Y., & Shleifer, A. (2016). A real estate boom with Chinese characteristics. *Journal of Economic Perspectives,* 31(1), 93–116.

Guo, W. (2019, December 25). Jin Yan'an: xibei zuida jinrushi minus jiqun [Golden Yan'an: The largest immersive B & B cluster in Northwest China]. *Zhongguo lvyou bao*, p. 3.

Harvey, D. (2003). *The new imperialism*. Oxford University Press.

Harvey, D. (2010a). *The enigma of capital*. Profile Books.

Harvey, D. (2010b). *A companion to Marx's capital*. Verso.

Herman, E. S., & Chomsky, N. (1988/2008). *Manufacturing consent: The political economy of the mass media*. Pantheon.

Huaao Yan'an shengdi hegu wenhua lvyou xintuo dianping [Comments on the trust between Sino-Australian International Trust and Yan'an Valley of Sacred Land cultural tourism]. (2014, March 25). *Sina*. Retrieved November 8, 2021 from http://finance.sina.com.cn/trust/20140325/152418607857.shtml

Huang, Y. (2008). *Capitalism with Chinese characteristics: Entrepreneurship and the state*. Cambridge University Press.

Jameson, F. (1991). *Postmodernism, or, the cultural logic of late capitalism*. Duke University Press.

Jameson, F. (2000). Globalization and political strategy. *New Left Review*, *4*, 49–68.

Jhally, S. (2000). Advertising at the edge of the apocalypse. In R. Andersen & L. Strate (Eds.), *Critical studies in media commercialism* (pp. 27–39). Oxford University Press.

Li, H., Meng, L., & Zhang, J. (2006). Why do entrepreneurs enter politics? Evidence from China. *Economic Inquiry*, *44*(3), 559–578.

Li, Q., & Li, S. (2014). Xinwen ren lanyong "ruan quanli" de weihai ji fangfan [Harm and prevention of newsmen abusing "soft power"]. *Qinghai jizhe*, *22*, 23–24.

Li, W. (2016). Failure by design–National mandates and agent control of local land use in China. *Land Use Policy*, *52*, 518–526.

Liu, N., & Chen, G. (2012, December 13). Wenshan quanguo shouchuang "zhenqiang shidan" hongse lvyou [Wenzhou businessman pioneers the use of "real gun and ammunition" in Red Tourism] *Wenzhou shangbao*, p. 24. Retrieved November 8, 2021 from http://news.qcc.com/postnews_a310316b8ff725c4485d70a8d15232c2.html

Long, H., Zou, J., & Liu, Y. (2009). Differentiation of rural development driven by industrialization and urbanization in eastern coastal China. *Habitat International*, *33*(4), 454–462.

MacKinnon, S. R., & Friesen, O. (Eds.). (1978). *China reporting*. University of California Press.

Marx, K. (1967). *Capital: A critique of political economy*. International Publishers.

McChesney, R. W. (2004). *The problem of the media*. Monthly Review Press.

McChesney, R. W. (2013). *Digital disconnect: How capitalism is turning the Internet against democracy*. New Press.

Mosco, V. (2009). *The political economy of communication*. Sage.

Naureckas, J. (2015, April 16). *They have "propaganda," US has "public diplomacy"– and a servile private sector*. Fair.org. http://fair.org/home/they-have-propaganda-us-has-public-diplomacy-and-a-servile-private-sector/

Palmeri, C., & Lin, L. (2014, April 29). Disney increases Shanghai Park investment to $5.5 billion. *Bloomberg News*. https://www.bloomberg.com/news/articles/2014-04-28/disney-increases-shanghai-park-investment-to-5-5-billion

Park, C. H. (2014). *Nongjiale* tourism and contested space in rural China. *Modern China*, *40*(5), 519–548.

Postman, N. (2000). Advertising at the edge of the apocalypse." In R. Andersen & L. Strate (Eds.), *Critical studies in media commercialism* (pp. 47–53). Oxford University Press.

Rubin, I. I. (1973). *Essays on Marx's theory of value*. Black Rose Books.

Sacred Land Valley Cultural Tourism Industrial Park Management Committee of Yan'an. (2014, December 11). *Yuanqu guihua* [Park planning]. http://www.yaholyvalley.com/index.php?m=content&c=index&a=show&catid=67&id=21

Sacred Land Valley Cultural Tourism Industrial Park Management Committee of Yan'an. (2016a, October 16). *Zhongxuanbu fu buzhang Jing Junhai diaoyan shengdi he gu jin Yan'an* [Deputy Minister of the Central Propaganda Department Jing Junhai visits the Valley and Golden Yan'an]. http://www.yaholyvalley.com/index.php?m=content&c=index&a=show&catid=41&id=259

Sacred Land Cultural Valley Tourism Industrial Park Management Committee of Yan'an. (2016b, August 1). *Zhongguo nongye yinhang yu shanlv jituan Yan'an gongsi juxing rongzi zuotan* [Agricultural Bank of China and STG Yan'an company held a financing talk]. http://www.yaholyvalley.com/index.php?m=content&c=index&a=show&catid=28&id=149

Sargeson, S. (2016). Grounds for self-government? Changes in land ownership and democratic participation in Chinese communities. *The Journal of Peasant Studies*, *45*(2), 321–346.

Shengdi hegu jin Yan'an [Golden Yan'an of the Sacred Land Valley]. (2021, May 8). http://www.yaholyvalley.com/index.php/index/index/about_detail.html?id=6&art_id=5

Song, Y., Wang, M. Y., & Lei, X. (2016). Following the money: Corruption, conflict, and the winners and losers of suburban land acquisition in China. *Geographical Research*, *54*(1), 86–102.

Statistical Bureau of Yan'an City. (2016). *2015 nian Yan'an guomin jingji he shehui fazhan tongji gongbao* [The statistical report of Yan'an city on the 2015 national economic and social development]. http://www.yanan.gov.cn/gk/tjxx/ndtj/267808.htm

STG. (n.d.). Shan lv gaikuang [Overview of STG] http://www.sxtourgroup.com/about.aspx?aid=187

Tian, G. (2014, May 12). Shengdi hegu xiangmu Zhengyi Zhong fugong hongse lvyou bian maidi youxi? [The Sacred Land Valley project restored in controversy, Red Tourism transforms into a game of land sale? *Zhongguo fangdichan bao*. Retrieved November 8, 2021 from https://www.163.com/money/article/9S4O51UD002534NU.html

Tsetsura, K. (2015). *Guanxi*, gift-giving, or bribery? Ethical considerations of paid news in China. *Public Relations Journal*, *9*(2), 1–26.

Wang, H. (2003). *China's new order: Society, politics, and economy in transition*. Harvard University Press.

Wang, W., & Ye, F. (2016). The political economy of land finance in China. *Public Budgeting & Finance*, *36*(2), 91–110.

Webber, M. (2012). *Making capitalism in rural China*. Edward Elgar.

Weber, M. (1958). *The Protestant ethic and the spirit of capitalism*. Scribner.

Wilmsen, B. (2016). Expanding capitalism in rural China through land acquisition and land reforms. *Journal of Contemporary China*, *25*(101), 1–17.

Xi, J., Wang, X., Kong, Q., & Zhang, N. (2015). Spatial morphology evolution of rural settlements induced by tourism. *Journal of Geographical Sciences*, *25*(4), 497–511.

Xu, H., Wu, Y., & Wall, G. (2012). Tourism real estate development as a policy tool for urban tourism: A case study of Dali and Lijiang, China. *Journal of China Tourism Research*, *8*(2), 174–193.

Xu, R. (2021, July 20). Jinxiang lishi, xieyi Yan'an [Mirroring history and capturing the soul of Yan'an]. *Yan'an ribao*, p. 7.

Yi, Y., & Yu, D. (2013). Yan'an shi jingdian hongse lvyou jingqu fazhan cunzai de wenti ji duice [The problems and countermeasures of the development of the classic Red Tourism scenic spots in Yan'an]. *Zhongwai qiyejia, 6*(428), 7–8.

Yici yiwai liaoshen zhanyi jingqu geng zhide qi dai [An accident makes the Liaoshen Campaign reenactment more riveting]. (2014, June 30). *Fushun wanbao*, p. A2.

Zan, Y. (2015, April 24). Fushun liaoshen zhanyi jingqu zhengshi kaiye [Liaoshen Campaign scenic spot opens in Fushun]. *Fushun ribao*, p. 6. http://daily.0245.cc/html /2015-04/24/node_7.htm

Zhao, S. N., & Timothy, D. J. (2015). Governance of red tourism in China: Perspectives on power and *guanxi*. *Tourism Management, 46*, 489–500.

Index

For Product Safety Concerns and Information please contact our EU
representative GPSR@taylorandfrancis.com
Taylor & Francis Verlag GmbH, Kaufingerstraße 24, 80331 München, Germany

www.ingramcontent.com/pod-product-compliance
Lightning Source LLC
Chambersburg PA
CBHW060303220326
41598CB00027B/4214